Herbert D. Gullick

CIRCLES
A MATHEMATICAL VIEW

D. PEDOE
B. Sc. (Lond.), B.A., Ph. D. (Cantab.)

Professor of Mathematics
University of Minnesota

Dover Publications, Inc.
New York

FOR NAOMI

Published in Canada by General Publishing Company, Ltd., 30 Lesmill Road, Don Mills, Toronto, Ontario.
Published in the United Kingdom by Constable and Company, Ltd., 10 Orange Street, London WC2H 7EG.

This Dover edition, first published in 1979, is a corrected and enlarged republication of the work originally published in 1957 by Pergamon Press.

International Standard Book Number: 0-486-63698-4
Library of Congress Catalog Card Number: 78-73522

Manufactured in the United States of America
Dover Publications, Inc.
180 Varick Street
New York, N.Y. 10014

PREFACE TO THE DOVER EDITION

IN THIS edition some minor corrections have been made, and problems added, with solutions, which may help the reader to test his grasp of the first three chapters. A few titles of books for further reading are also suggested.

All mathematical works have defects, and the author is only too conscious of those in his book, but he believes that it is best to leave it as it was written, in the winter sunshine of Khartoum, and he hopes that it will continue to appeal to geometry lovers everywhere.

D. PEDOE

School of Mathematics
University of Minnesota
Minneapolis, Minn. 55455

PREFACE TO THE FIRST EDITION

THIS book presents some branches of pure mathematics in which a figure of universal appeal, the Circle, is the leading character. We begin in chapter I with some well-known properties of the circle, but preparations are also made for later chapters, and a number of classical problems solved. Chapter II deals with some of the interesting consequences of representing a circle by means of a point in ordinary space. Very little three-dimensional geometry needs to be known for an understanding of this chapter.

A celebrated model of non-Euclidean geometry is discussed in some detail in chapter III, and circles play a fundamental part again. Here also, the discussion is elementary. Chapter IV, perhaps the most difficult in this book, deals with the isoperimetric property of the circle, and necessarily uses some classical theorems from the theory of functions. But procedures which are not standard are described with care, and, in any case, are reserved for the end of the chapter, so that the ideas involved can be understood even by a beginner.

It is hoped that this book will appeal to those who deprecate the tendency to overspecialization in mathematics. It has arisen out of a series of intercollegiate lectures given when the author was Reader in Mathematics in the University of London. These lectures were designed for students in their first or second year. The reception then accorded to him has encouraged the author to hope that, in book form, his discourse on circles may also appeal to a wider audience.

D. PEDOE

CONTENTS

vii

CHAPTER III

CHAPTER IV

BOOKS FOR FURTHER READING

Guggenheimer, H. W. *Plane Geometry and Its Groups.* San Francisco: Holden-Day, 1967.

Pedoe, D. *A Course of Geometry for Colleges and Universities.* London and New York: Cambridge University Press, 1970.

———. *The Gentle Art of Mathematics.* New York: Dover Publications, 1973.

———. *Geometry and the Liberal Arts.* New York: St. Martin's Press, 1978.

CHAPTER I

THE properties of circles discussed in this chapter are those which have the habit of appearing in other branches of mathematics.

1. The nine-point circle

This circle is the first really exciting one to appear in any course on elementary geometry. It is a circle which is found to pass through nine points intimately connected with a given triangle ABC. Three proofs are given. Each has its own peculiar merits.

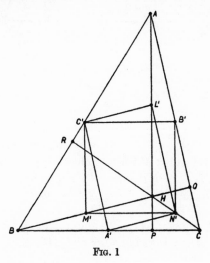

FIG. 1

(i) H is the orthocentre of triangle ABC, the intersection of the altitudes AP, BQ, CR. The midpoints of BC, CA and AB are A', B' and C' respectively. L', M', N' are the midpoints of HA, HB, HC. We show that a circle passes through the nine points A', B', C', L', M', N', P, Q, R. We use the theorem that the line joining the midpoints of two sides of a triangle is parallel to the third side.

Both $B'C'$ and $N'M'$ are $\parallel$ to BC, and both $B'N'$ and $C'M'$ are $\parallel$ to AH. Hence $B'C'M'N'$ is a rectangle. Similarly $C'A'N'L'$ is a rectangle.

1

Hence $A'L'$, $B'M'$ and $C'N'$ are three diameters of one circle. Since $A'P$ and $L'P$ are $\perp$, this circle passes through P. Similarly it passes through Q and R.

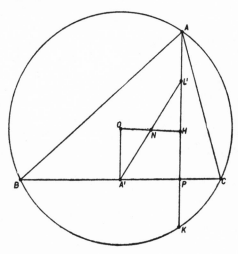

FIG. 2

(ii) Let O be the circumcentre of ABC and R the circumradius. We know that $AH = 2OA'$. Since L' bisects AH, $L'H = OA'$, and therefore OH and $L'A'$ bisect each other, in N, say.

Now $NL' = NA' = NP$ (since $\angle APB = 90°$). We also know that $HP = PK$, and since $ON = NH$, we have

$$NP = \tfrac{1}{2}OK = \tfrac{1}{2}R.$$

Therefore the circle centre N and radius $\tfrac{1}{2}R$ passes through P, A', L'. Similarly it passes through Q, B', M' and R, C', N'.

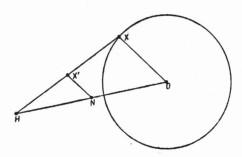

FIG. 3

(iii) The third proof uses a lemma from the theory of similar figures. If X is a point on a given circle of centre O and radius R, and H is any given fixed point, the locus of X', the midpoint of HX, is a circle centre N, where N bisects HO, and radius $\frac{1}{2}R$. The proof is clear.

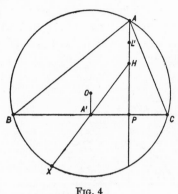

Fig. 4

Applying this lemma to the case of the circumcircle of ABC, H being the orthocentre, we see that a circle centre N and radius $\frac{1}{2}R$ passes through P, Q, R, L', M', N'. We must now show that this circle also passes through A', B', C'. It is sufficient to show that if HX contains A', then HX is bisected at A'.

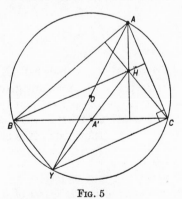

Fig. 5

Let AO meet the circumcircle in Y. Then $\angle ACY = 90°$, so that $YC \parallel BH$. Since also $\angle ABY = 90°$, $YB \parallel CH$. Therefore $YBHC$ is a parallelogram, and the diagonals YH and BC

bisect each other.   Hence Y is the same point as X in the previous diagram, and the theorem is proved.

The methods used above are elementary, and at least one of the three proofs will be familiar to most readers of this book.   In chapter II we shall need to assume some theorems in inversion and coaxal systems of circles.   As fashions in mathematical teaching change, it cannot be assumed that all readers will be familiar with these theories, and a brief (but adequate) outline of each will now be given.

2. Inversion

This is a one-to-one transformation of the points of the plane by

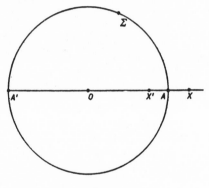

FIG. 6

means of a given circle, which we shall call Σ, of radius k and centre O.   To obtain the transform X' of any given point X, or the *inverse* of X in the circle Σ, we join X to O, and find X' on OX such that $OX.OX' = k^2$.   The point O itself is excluded from the points of the plane which may be transformed.   It is clear that

(i) X is the inverse of X';
(ii) points on Σ transform into themselves;
(iii) if A, A' are the ends of the diameter of Σ through X, then X, X' are harmonic conjugates with respect to A, A'.

All circles through X and X' cut Σ orthogonally, since the square of the tangent from O to such a circle $= OX.OX' = k^2$.

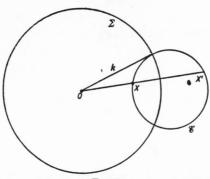

Fig. 7

As a corollary to this remark, we see that if $\mathscr{C}$ is any circle orthogonal to Σ, all points on $\mathscr{C}$ invert into points on $\mathscr{C}$, for $OX.OX' = k^2$. It is sometimes necessary to consider the process of inversion when Σ, the circle of inversion, is a line. If we keep A fixed, and let A' move off to infinity, we see that X' moves towards the geometrical image of X in the resulting line.

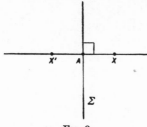

Fig. 8

For future application we now give a construction, using only a

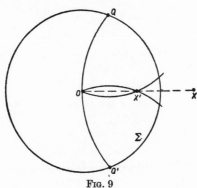

Fig. 9

pair of compasses, by which we can find the inverse of a point X in a circle Σ of given centre O. Let the circle centre X and radius XO cut Σ in Q and Q', and let O and X' be the intersections of the two circles with centres Q and Q' and radii QO and $Q'O$ respectively. Then X' is the inverse of X in Σ.

For QOX and $X'OQ$ are isosceles triangles with a common base angle. They are therefore similar, and $OX : OQ = OQ : OX'$. Hence $OX . OX' = OQ^2$.

This construction lies at the basis of the discussion on *compass geometry* in §11.

We are interested in the locus of inverse points X' as X describes a given curve $\mathscr{C}$. This locus is called the *inverse of the curve* $\mathscr{C}$ with respect to Σ. When $\mathscr{C}$ is a circle orthogonal to Σ, we have seen that the inverse is $\mathscr{C}$ itself. The general theorem is:

The inverse of a circle is a straight line or a circle.

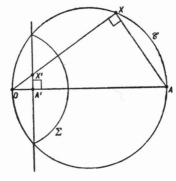

FIG. 10

(i) The centre of inversion O is on $\mathscr{C}$:

Let OA be the diameter of $\mathscr{C}$ through O, and let A' be the inverse of A. Then $OA . OA' = k^2$. If X is any point on $\mathscr{C}$, and X' its inverse,

$$OX . OX' = k^2 = OA . OA'.$$

Hence A, X, X', A' are concyclic points, and since $\angle AXO = 90°$, $\angle X'A'A = 90°$. Therefore X' describes the line through $A' \perp$ to OA.

We note that this line is $\parallel$ to the tangent to $\mathscr{C}$ at O. It naturally passes through the intersections of $\mathscr{C}$ and Σ, since points on Σ invert into themselves.

Corollary. The inverse of a straight line is a circle through the centre of inversion.

(ii) The centre of inversion O is not on $\mathscr{C}$. Let C be the centre

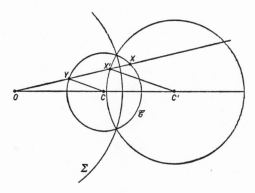

Fig. 11

of $\mathscr{C}$. Since

$$OX . OX' = k^2,$$

and

$$OX . OY = t^2,$$

where t is the tangent[†] from O to $\mathscr{C}$, by division we have

$$\frac{OX'}{OY} = \frac{k^2}{t^2}.$$

Since this ratio is constant, it follows that if C' is a point on OC such that $C'X' \parallel CY$, then C' is a fixed point, and $C'X'$ is constant. Hence X' moves on a circle centre C'.

Note that C' is not necessarily the inverse of C. We shall soon find what point the centre of $\mathscr{C}$ does invert into.

We need only one further theorem from the theory of inversion:

The angle of intersection of two circles is unaltered by inversion.

In fact a more general theorem is true, but this one will suffice for our subsequent needs.

[†] We shall see later that t^2 may be negative. Then O is inside $\mathscr{C}$.

(i) If both circles $\mathscr{C}$ and $\mathscr{D}$ pass through the centre O of inversion, they transform into straight lines parallel to the tangents to

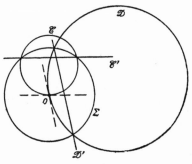

Fig. 12

$\mathscr{C}$ and $\mathscr{D}$ at O, so that the theorem is evident.

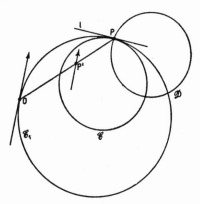

Fig. 13

(ii) If at least one of the circles $\mathscr{C}$, $\mathscr{D}$ does not pass through O, let P be a point of intersection of the two circles, and let l, m be the tangents at P to $\mathscr{C}$, $\mathscr{D}$ respectively. We may draw a circle $\mathscr{C}_1$ through O which touches l at P, and a circle $\mathscr{D}_1$ through O which touches m at P. Then the angle between $\mathscr{C}_1$ and $\mathscr{D}_1$ is the same as that between $\mathscr{C}$ and $\mathscr{D}$.

Now if P' is the inverse of P, the tangent at P' to the inverse of $\mathscr{C}$ is the inverse of $\mathscr{C}_1$. Hence the angle between the circles inverse to $\mathscr{C}$ and to $\mathscr{D}$ is equal to the angle between $\mathscr{C}_1$ and $\mathscr{D}_1$ (by (i) above), and this is equal to the angle between $\mathscr{C}$ and $\mathscr{D}$.

With the help of this theorem we can now find the inverse of the centre of a circle $\mathscr{C}$. All lines through the centre C cut $\mathscr{C}$ orthogonally. These lines invert into circles through the centre of inversion O and the inverse of C. Since all circles through O and the inverse of C cut the inverse of $\mathscr{C}$ orthogonally, the inverse of C must be *the inverse of O in the inverse of $\mathscr{C}$.*

If O lies on $\mathscr{C}$, so that the inverse of $\mathscr{C}$ is a straight line, the centre of $\mathscr{C}$ inverts into the geometrical image of O in this line.

We now give some applications of the theory of inversion.

3. Feuerbach's theorem

The nine-point circle of a triangle touches the incircle and excircles of the triangle.

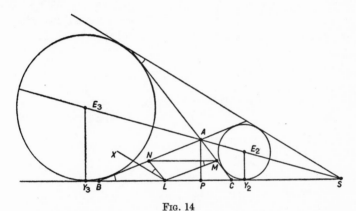

Fig. 14

We consider the excircles opposite B and C. Let their centres be E_2 and E_3, and their points of contact with BC Y_2 and Y_3. The points A, S divide E_2E_3 internally and externally in the same ratio, that of the radii of the two excircles. Dropping perpendiculars on BC, it follows that P and S also divide Y_2Y_3 internally and externally in the same ratio. That is, P and S are harmonic conjugates with respect to Y_2 and Y_3.

Let L, M, N be midpoints of BC, CA, AB. We note that $BY_2 = \frac{1}{2}(BC + CA + AB) = CY_3$, so that L is also the midpoint of Y_2Y_3. From the harmonic relation just proved we see that

$$LP \cdot LS = LY_2{}^2 = LY_3{}^2.$$

Now let LX be the tangent at L to the circle LMN, where X is on the same side of BC as N. Then
$$\angle XLN = \angle NML = \angle ABC.$$
Hence the angle between LX and $CA = \angle ABC$. It follows that LX is $\parallel$ to the fourth common tangent of the two excircles, the common tangent through S other than BC.

Now since $LP.LS = LY_2{}^2 = LY_3{}^2$, the nine-point circle LMN is the inverse of this fourth common tangent with respect to the circle centre L and radius LY_2. For this nine-point circle contains P, and inverts into the line through S parallel to the tangent to the circle at L.

Also since the circle of inversion cuts both excircles orthogonally, each excircle inverts into itself. We have seen that the inverse of the nine-point circle touches these excircles. It follows that the nine-point circle touches the excircles centres E_2, E_3. In precisely the same way we can show that the nine-point circle touches the incircle and the excircle opposite A.

4. Extension of Ptolemy's theorem

If A, B, C, D are any four points in a plane, then
$$AB.CD + AD.BC > AC.BD,$$
unless A, B, C, D lie, in the order $ABCD$, on a circle or a straight line. In the latter case, the inequality becomes an equality.

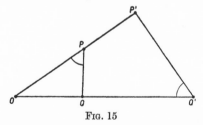

FIG. 15

This is proved by noting the influence of inversion on the length of a segment. If the points P, Q invert into the points P', Q', with O as centre of inversion, then OPQ and $OQ'P'$ are similar triangles. Hence
$$\frac{P'Q'}{PQ} = \frac{OP'}{OQ} = \frac{OP.OP'}{OP.OQ} = \frac{k^2}{OP.OQ},$$
so that
$$P'Q' = \frac{k^2\,PQ}{OP.OQ}.$$

Now, given the four points A, B, C, D, we invert with respect to A. Let B', C', D' be the respective inverses of B, C, D. Then $B'C' + C'D' > B'D'$ unless C' lies on the line $B'D'$ between B' and D'. In the latter case we have

$$B'C' + C'D' = B'D'.$$

The first inequality becomes

$$\frac{BC}{AB.AC} + \frac{CD}{AC.AD} > \frac{BD}{AB.AD},$$

or

$$AB.CD + AD.BC > AC.BD,$$

unless C' lies on the line $B'D'$ between B' and D', in which case A, B, C, D lie on a circle in the order $ABCD$, or on a straight line.

5. Fermat's problem

A, B, C are any three points in a plane.   To find a point P such that $PA + PB + PC$ shall be least.

Let B and C be the acute angles of the triangle ABC.   On BC, and away from A, describe an equilateral triangle BCD.   Then by the extension of Ptolemy's theorem, unless P lies on the circle

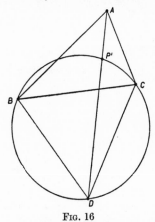

Fig. 16

BCD so that the order of the points is $BPCD$, we have

$$BP.CD + PC.DB > PD.BC,$$

or

$$PB + PC > PD,$$

since $CD = DB = BC$.   Therefore

$$PA + PB + PC > PA + PD.$$

Now, unless P lies on AD, we have $PA + PD > AD$. Hence, unless P is at P' (the other intersection of AD with the circle BCD), we have $PA + PB + PC > AD$.

But if P is at P', both the above inequalities become equalities; so that $P'A + P'B + P'C = AD$.

Hence $P'A + P'B + P'C < PA + PB + PC$.

Therefore P' is the required point.

If $\angle BAC = 120°$, $A = P'$, and A is the required point.

If $\angle BAC > 120°$, A is still the required point.

6. The centres of similitude of two circles

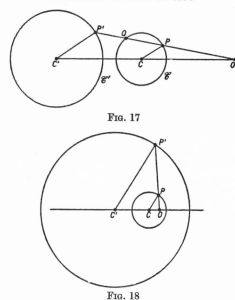

FIG. 17

FIG. 18

These points have already appeared in some of our proofs. We now define them.

External centre of similitude. Let $\mathscr{C}$ and $\mathscr{C}'$ be two circles of unequal radii, with centres C and C'. Let CP and $C'P'$ be parallel radii drawn in the *same* sense. Then it is clear that $P'P$ cuts $C'C$ in a fixed point O, which divides $C'C$ externally in the ratio of the radii of the two circles. This is the external centre of similitude. If the ratio of the radii $= k$, then $OP' : OP = k$. We also have $OP \cdot OQ = $ constant. Therefore

$$OP' \cdot OQ = \text{constant.}$$

It follows that with centre of inversion O, and a suitable radius of inversion, $\mathscr{C}$ can be inverted into $\mathscr{C}'$.

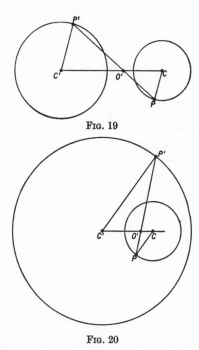

FIG. 19

FIG. 20

Internal centre of similitude. Here the radii of the circles $\mathscr{C}$ and $\mathscr{C}'$ may be equal. CP and $C'P'$ are parallel radii in opposite senses. $P'P$ cuts $C'C$ in a fixed point O', the internal centre of similitude, which divides $C'C$ internally in the ratio of the radii.

Direct common tangents of $\mathscr{C}$ and $\mathscr{C}'$, if they exist, cut the line of centres in O, and transverse common tangents, if they exist, cut the line of centres in O'.

If $\mathscr{C}$ and $\mathscr{C}'$ are concentric, the common centre is both the internal and external centre of similitude.

If $\mathscr{C}$ and $\mathscr{C}'$ touch externally, that is, if they lie on opposite sides of the tangent at the point of contact, this point is the *internal* centre of similitude. If the circles touch internally, the point of contact is the *external* centre.

If we have three circles, no two of which have equal radii, and the centres of which form a triangle, the six centres of similitude lie in threes on four straight lines. This is easily proved, using the theorem of Menelaus.

7. Coaxal systems of circles

These are best introduced by using the methods of coordinate geometry.

The equation of any line passing through the origin of coordinates in the (x,y)-plane may be written as

$$\lambda x + \mu y = 0.$$

These lines are said to form a *pencil* of lines, and the equation of the system depends *linearly* on two distinct lines of the pencil, given by $x = 0$ and $y = 0$.

Similarly, if $u \equiv ax + by + c = 0$ and $v \equiv a'x + b'y + c' = 0$ are any two distinct lines, the line

$$\lambda u + \mu v = 0$$

represents, for suitable λ, μ, any line through the intersection of $u = 0$ and $v = 0$.

When we consider two distinct circles $\mathscr{C}$ and $\mathscr{C}'$, given by the equations

$$C \equiv x^2 + y^2 + 2gx + 2fy + c = 0,$$
$$C' \equiv x^2 + y^2 + 2g'x + 2f'y + c' = 0,$$

the system $\lambda C + \mu C' = 0$, derived linearly from them, is a system of circles, since it may also be written as

$$x^2 + y^2 + \frac{2x(\lambda g + \mu g')}{\lambda + \mu} + \frac{2y(\lambda f + \mu f')}{\lambda + \mu} + \frac{\lambda c + \mu c'}{\lambda + \mu} = 0.$$

If (x',y') is a point of intersection of $\mathscr{C}$ and $\mathscr{C}'$,

$$C(x',y') = C'(x',y') = 0,$$

that is, the point satisfies the equations of both circles. Since also, for all values of λ, μ,

$$\lambda C(x',y') + \mu C'(x',y') = 0,$$

this same point lies on all circles of the system $\lambda C + \mu C' = 0$. Such a system of circles therefore passes through the points of intersection (if any) of $\mathscr{C}$ and $\mathscr{C}'$. The system may be described as a *pencil* of circles, but for reasons which will soon become clear, we give it the more normal title of a *coaxal system* of circles.

A geometrical meaning may be given to the ratio of the parameters $\lambda : \mu$. Let $P(x',y')$ be any point in the plane, and let any line through P cut $\mathscr{C}$ in Q and R. Then it is elementary that the product

$$PQ.PR = x'^2 + y'^2 + 2gx' + 2fy' + c.$$

This constant, which is > 0 if P is outside the circle and equal to the square of the tangent from P to $\mathscr{C}$, is called, in all cases, the *power* of P with respect to $\mathscr{C}$. If P is inside $\mathscr{C}$, the power is negative, and it is zero, of course, if P is on $\mathscr{C}$. Since

$$\lambda C + \mu C' = 0$$

may be written as

$$\frac{\lambda}{\mu} = \frac{-C'}{C},$$

we see that *the locus of a point which moves so that the ratio of its powers with respect to two given circles $\mathscr{C}$ and $\mathscr{C}'$ is a constant is a circle of the coaxal system determined by $\mathscr{C}$ and $\mathscr{C}'$*.

A particular and important member of the system is obtained by taking the ratio of the powers to be unity. We then obtain

$$C - C' \equiv 2(g - g')x + 2(f - f')y + c - c' = 0,$$

a straight line, called the *radical axis* of the system. If $\mathscr{C}$ and $\mathscr{C}'$ intersect, this line is the common chord of the two circles, and it is evident that points on the common chord of two circles do have equal powers with respect to each circle. But the radical axis

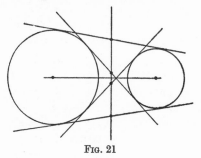

FIG. 21

exists even if the circles do not intersect. It is easily verified that the radical axis of two circles is perpendicular to the line of centres.

We deduce that the midpoints of the common tangents of two circles lie on a line which is perpendicular to the line of centres.

We obtain the same coaxal system of circles if we take any two distinct members of the system $\lambda C + \mu C' = 0$, say

$$C_1 \equiv \lambda_1 C + \mu_1 C' = 0,$$

and

$$C_2 \equiv \lambda_2 C + \mu_2 C' = 0,$$

where $\lambda_1/\mu_1 \neq \lambda_2/\mu_2$, and write down the system of circles which depend linearly on C_1 and C_2. This should be verified. Another theorem of importance is that any two circles of a coaxal system have the same radical axis. This explains the term *coaxal system*. We can verify this directly. The radical axis of C_1 and C_2 is

$$2x\left[\frac{\lambda_1 g + \mu_1 g'}{\lambda_1 + \mu_1} - \frac{\lambda_2 g + \mu_2 g'}{\lambda_2 + \mu_2}\right] + 2y\left[\frac{\lambda_1 f + \mu_1 f'}{\lambda_1 + \mu_1} - \frac{\lambda_2 f + \mu_2 f'}{\lambda_2 + \mu_2}\right]$$
$$+ \left[\frac{\lambda_1 c + \mu_1 c'}{\lambda_1 + \mu_1} - \frac{\lambda_2 c + \mu_2 c'}{\lambda_2 + \mu_2}\right] = 0.$$

This easily reduces to

$$(\lambda_1\mu_2 - \lambda_2\mu_1)[2x(g - g') + 2y(f - f') + c - c'] = 0.$$

It follows that the centres of all circles of a coaxal system lie on a line perpendicular to the radical axis. This is also clear from the fact that the centre of $\lambda C + \mu C' = 0$ is the point

$$\left[\frac{\lambda(-g) + \mu(-g')}{\lambda + \mu}, \quad \frac{\lambda(-f) + \mu(-f')}{\lambda + \mu}\right].$$

In the next chapter we shall need to be familiar with a few more properties of coaxal systems, and we derive them here.

8. Canonical form for coaxal system

The line of centres and the radical axis of a coaxal system offer, in a sense, a natural system of coordinate axes for the system. We take the line of centres as the x-axis, and the radical axis as the y-axis, and see what form the equation of the system takes.

Since their centres are on the x-axis, any two circles $\mathscr{C}$ and $\mathscr{C}'$ have equations of the form,

$$C \equiv x^2 + y^2 + 2gx + c = 0,$$

and

$$C' \equiv x^2 + y^2 + 2g'x + c' = 0.$$

The radical axis of these two circles is

$$C - C' \equiv 2(g - g')x + c - c' = 0,$$

and since this is to be $x = 0$, we must have $c = c'$. The circles of the system may therefore be written in the form

$$x^2 + y^2 + 2\lambda x + c = 0,$$

where c is constant, and λ a parameter. This is a canonical, or standard form for a coaxal system.

We note that the system now depends linearly on the circle $x^2 + y^2 + c = 0$, and the radical axis $x = 0$.

We can now easily distinguish between the intersecting and non-intersecting type of coaxal system. Intersections, if any, must lie on the radical axis $x = 0$. This meets circles of the system where $y^2 + c = 0$. If c is positive, there are no intersections, and we have the non-intersecting type of coaxal system. If c is negative, all circles of the system cut the radical axis in the same two points.

(i) $c = k^2 > 0$. Any circle of the system may be written in the form

$$(x + \lambda)^2 + y^2 = \lambda^2 - k^2.$$

For real circles, we must have $\lambda^2 \geqslant k^2$. There are two circles of zero radius in the system, with centres at $(-k, 0)$ and $(k, 0)$. These points, which we denote by L and L', are called the *limiting*

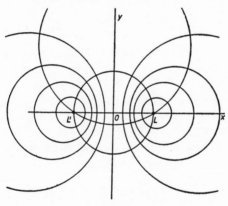

Fig. 22

points of the system. No circles of the system have their centres between L and L'. If a circle of the system cuts the line of centres in A, A', then $OA . OA' = k^2 = OL^2$, and it follows that L and L' are harmonic conjugates with respect to A and A'. Hence any circle through L and L' cuts all circles of the coaxal system orthogonally.

(ii) $c = -k^2 < 0$. This case presents no features which are

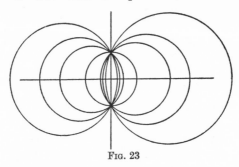

Fig. 23

not evident geometrically.

(iii) $c = 0$. This can be regarded as an intermediate case.

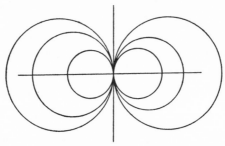

Fig. 24

All circles of the system touch the radical axis, and the point of contact may be regarded as consisting of two coincident limiting points.

We have remarked that in (i) any circle through L and L' is orthogonal to every circle of the given system. The equation of a circle through L and L' is

$$x^2 + y^2 + 2\mu y - k^2 = 0,$$

and this, of course, is a coaxal system of the intersecting type. The orthogonality relation is evident algebraically, since the condition to be satisfied, $2gg' + 2ff' - c - c' = 0$, becomes here $2\lambda.0 + 2.0.\mu - k^2 + k^2 = 0$, which is the case.

Hence, associated with every coaxal system of a given type there is a coaxal system of the opposite type, and the circles of one cut the circles of the other orthogonally. Such systems are called *conjugate systems*, for reasons which will become clear in chapter II.

9. Further properties

If we try to find the circles of the coaxal system

$$x^2 + y^2 + 2\lambda x + c = 0$$

which pass through $P(x',y')$, we see that

$$\lambda = -\frac{(x'^2 + y'^2 + c)}{2x'} \; ;$$

that is, there is a unique value of the parameter which determines the circle. Hence one circle of a coaxal system passes through a given point P. The construction of this circle is evident for intersecting systems, since a circle is determined by three points, but not so evident for a non-intersecting system. In this case we invoke the *conjugate system*, which is of the intersecting type, and draw the unique circle of this system which passes through P. We can then construct the required circle by drawing the unique circle through P which is orthogonal to the constructed circle and has its centre on the line of centres of the given coaxal system.

If we invert a coaxal system of circles, it is fairly clear that we obtain an intersecting coaxal system in the intersecting case, but this is not so clear in the non-intersecting case. However we can bring in the *conjugate system*, which is of the intersecting type, and remembering that circles which are orthogonal invert into orthogonal circles, we see that the circles of the non-intersecting system invert into circles which are orthogonal to all the circles of a coaxal system. Such a system of circles must form the conjugate coaxal system.

However, inversion is a transformation which sometimes does more than we demand of it, and the reader may feel reluctant to accept the proof indicated when we show, as we do now, that *any non-intersecting system of coaxal circles may be inverted into a system of concentric circles.* The question he will ask is: if a system of concentric circles is a coaxal system, why have we not come across it in the above treatment?

We leave the reader to ponder on this apparent paradox, and prove the theorem. Let L and L' be the limiting points of the non-intersecting coaxal system, and invert with respect to a circle centre L. Then the circles of the system invert into circles which are orthogonal to the inverses of all circles through L and L'. But these inverses are straight lines through a fixed point, the inverse of L'. Hence the coaxal system inverts into a system of concentric circles, with centre at the inverse of L'.

This theorem provides a solution of the attractive problem:

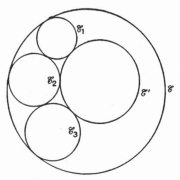

FIG. 25

given two circles $\mathscr{C}$ and $\mathscr{C}'$, one lying inside the other, can we find a chain of circles, $\mathscr{C}_1$, $\mathscr{C}_2$, $\mathscr{C}_3$, ..., $\mathscr{C}_n$, all of which touch $\mathscr{C}$ and $\mathscr{C}'$, such that $\mathscr{C}_1$ touches $\mathscr{C}_2$ and $\mathscr{C}_n$, $\mathscr{C}_2$ touches $\mathscr{C}_3$ and $\mathscr{C}_1$, ..., $\mathscr{C}_n$ touches $\mathscr{C}_{n-1}$ and $\mathscr{C}_1$?

Inverting $\mathscr{C}$ and $\mathscr{C}'$ into concentric circles, we see that the

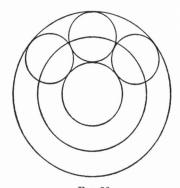

FIG. 26

problem does not always have a solution, but that when there is a solution, we can begin the chain by drawing any circle which touches $\mathscr{C}$ and $\mathscr{C}'$, and can then fit in the other members of the chain in an obvious manner.

We conclude this discussion of coaxal circles by considering the radical axes, in pairs, of three circles $\mathscr{C}$, $\mathscr{C}'$, $\mathscr{C}''$. The intersection of two of these radical axes is a point which has equal powers with respect to all three circles. This point is therefore on the third radical axis. Hence, if they are not parallel, the radical axes of the three pairs of circles which can be formed from three given circles are concurrent. The point of concurrence is called the *radical centre* of the three circles. The circle which has its centre at the radical centre, and radius equal to the length of the tangent from the centre to any one of the three circles, if it lies outside one, and therefore all three circles, cuts each of the three circles orthogonally and is the only circle which does this. With respect to this circle as circle of inversion, each of the three circles inverts into itself.

10. Problem of Apollonius

Given three circles $\mathscr{C}_1$, $\mathscr{C}_2$ and $\mathscr{C}_3$, to draw a circle touching all three.

We need two preliminary results:

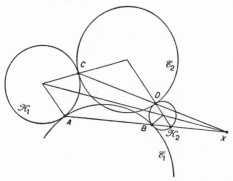

Fig. 27

(i) If two non-concentric circles $\mathscr{K}_1$, $\mathscr{K}_2$ touch two other non-concentric circles $\mathscr{C}_1$, $\mathscr{C}_2$ at A, B and C, D respectively, and if the contacts at C and D are of like[†] or unlike type according as those at A and B are, then AB and CD meet in a centre of similitude of the circles $\mathscr{K}_1$, $\mathscr{K}_2$, or they are parallel to the line of centres of these circles.

To prove this result we join the centres of $\mathscr{K}_1$, $\mathscr{K}_2$ and $\mathscr{C}_1$ and find, using the theorem of Menelaus, where AB cuts the line joining the centres of $\mathscr{K}_1$, $\mathscr{K}_2$. If the given conditions are satisfied, and we consider the triangle formed by the centres of $\mathscr{K}_1$,

---

† Contacts are of " like type " when both are external or both are internal.

$\mathscr{K}_2$ and $\mathscr{C}_2$, we see that CD passes through the same point X, and that this is a centre of similitude of $\mathscr{K}_1$ and $\mathscr{K}_2$.

(ii) We remarked earlier (see page 12) that $XA.XB = XC.XD$, so that X is on the radical axis of $\mathscr{C}_1$ and $\mathscr{C}_2$.

We also note that if the contacts at A and B are of like or unlike type according as those at C and D are, then the contacts at A and C are of like or unlike type according as those at B and D are. Therefore AC and BD meet at a centre of similitude of $\mathscr{C}_1$ and $\mathscr{C}_2$, and this centre is on the radical axis of $\mathscr{K}_1, \mathscr{K}_2$.

We have therefore shown that the radical axis of $\mathscr{C}_1$ and $\mathscr{C}_2$ contains a centre of similitude of $\mathscr{K}_1$ and $\mathscr{K}_2$, and the radical axis of $\mathscr{K}_1$ and $\mathscr{K}_2$ contains a centre of similitude of $\mathscr{C}_1$ and $\mathscr{C}_2$.

We now return to our problem, and assume, for simplicity, that the radii of $\mathscr{C}_1$, $\mathscr{C}_2$, $\mathscr{C}_3$ are three distinct numbers, so that all the centres of similitude exist.

Let $\mathscr{K}_0$ be a circle touching the three circles externally. Inver-

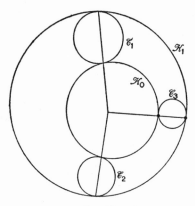

Fig. 28

sion from the radical centre of $\mathscr{C}_1$, $\mathscr{C}_2$, $\mathscr{C}_3$ changes these circles into themselves, and $\mathscr{K}_0$ is transformed into a circle $\mathscr{K}_1$ which is touched internally by the three circles. The joins of the points of contact of each circle $\mathscr{C}_i$ $(i = 1, 2, 3)$ with $\mathscr{K}_0$ and with $\mathscr{K}_1$ pass through the radical centre.

The pole, for $\mathscr{C}_1$, of the line joining the points of contact of $\mathscr{K}_0$ and $\mathscr{C}_1$ and of $\mathscr{K}_1$ and $\mathscr{C}_1$ is the intersection of the tangents at these points of contact. These tangents are the radical axes of $\mathscr{K}_0$ and $\mathscr{C}_1$ and of $\mathscr{K}_1$ and $\mathscr{C}_1$, and since the radical axes of three circles, taken in pairs, are concurrent, this pole lies on the radical axis of $\mathscr{K}_0$ and $\mathscr{K}_1$.

By the reciprocal property of pole and polar, the polar of this point contains the pole, for $\mathscr{C}_1$, of the radical axis. Now the polar is the line joining the points of contact of $\mathscr{K}_0$ and $\mathscr{C}_1$ and of $\mathscr{K}_1$ and $\mathscr{C}_1$ and passes through the radical centre of $\mathscr{C}_1$, $\mathscr{C}_2$, $\mathscr{C}_3$. Since also the radical axis of $\mathscr{K}_0$ and $\mathscr{K}_1$ contains the external centres of similitude of the circles $\mathscr{C}_1$, $\mathscr{C}_2$, $\mathscr{C}_3$ taken in pairs (this was proved in (ii) above), we have the following construction, due to Gergonne:

Draw the line through the external centres of similitude of $\mathscr{C}_1$, $\mathscr{C}_2$ and $\mathscr{C}_3$, taken in pairs, and join its poles for each of these circles to the radical centre of the three circles. These joins cut the given circles in points of contact of two of the required circles.

The centres of similitude of $\mathscr{C}_1$, $\mathscr{C}_2$, $\mathscr{C}_3$ lie by threes on four lines, and each line of centres of similitude gives two tangent circles. In the most favourable case we therefore expect to find *eight* solutions of the problem of Apollonius.

We shall verify this from another point of view in chapter II.

11. Compass geometry

We conclude this chapter by showing how the straight edge may be dispensed with in the elementary constructions of Euclidean geometry.

A straight edge is used to connect two given points by a straight line. No pair of compasses will do this. At the same time the straight edge and compasses are primarily used to determine certain points. The fundamental problems in Euclidean geometry are:

(i) to find the intersections of two circles, the centre and radius of each being given;

(ii) to find the intersection of a line given by two points with a circle given by its centre and radius;

(iii) to find the intersections of two lines, each given by two points.

The solution of (i) is naturally possible using only the compasses. Problems (ii) and (iii) reduce to (i) after inversion.

Therefore, to show that all Euclidean constructions are possible using only a pair of compasses, we must show:

(a) how to find the inverse of a given point in a circle of given centre and radius;

(b) how to find the centre of a circle which is to pass through three given points.

We gave the solution of (a) in §2, page 5.

Before dealing with (b), we show how to find, using compasses only, the midpoint of the segment determined by two given points. In the first place we show how to extend a given segment AO to

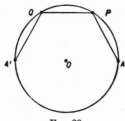

Fig. 29

twice its length. Draw the circle centre O and radius OA, and mark off the chords $AP = PQ = QA' = OA$. Then AOA' is a straight line, and $AA' = 2OA$.

We can now find the midpoint of the segment OA. Extend AO to A', where $AO = OA'$, and find the inverse of A' in the

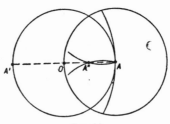

Fig. 30

circle centre A and radius AO. If A'' is the inverse,
$$AA'' \cdot AA' = AO^2,$$
and since $AA' = 2AO$, the point A'' is the midpoint of OA.

We can extend OA to A', where $OA' = nOA$, n being any positive integer, and the inverse of A' in the circle centre A and radius AO gives OA/n.

(b) Let A, B, C be the three given points. Invert with respect to the circle centre A and radius AB. The circle ABC inverts

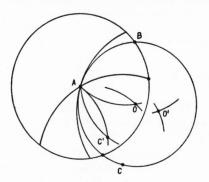

FIG. 31

into the line BC', where C' is the inverse of C in the circle of inversion. This point can be constructed. The centre of circle ABC inverts into the inverse of the centre of inversion, which is A, in the inverse of circle ABC, namely, the line BC' (see §2, page 9). The inverse of A in BC' is the geometrical image of A in BC'. This is easily constructed, as indicated in Fig. 31. If this geometrical image be O', we finally find the inverse of O' in the circle of inversion. This final point O is the centre of circle ABC.

The reader should carry out this construction for himself.

To find where two lines determined by points A,B and C,D respectively meet, we invert with respect to any suitable circle, centre O, say. If the respective inverse points are A',B' and C',D', the intersection of the circles $OA'B'$ and $OC'D'$, other than O, is the inverse of the point required. We can draw these circles, by the above construction, and find this point.

Bright students often rediscover this compass geometry for themselves, and are dashed to find that it was fully investigated, without the benefit of inversion, by Mascheroni in *La Geometria del Compasso*, Pavia, 1797.

CHAPTER II

1. Representation of a circle

In this chapter we adopt a completely different point of view from that pursued in the first chapter. We note that the equation of a circle is

$$C \equiv x^2 + y^2 + 2gx + 2fy + c = 0, \qquad (1)$$

and that this equation contains three parameters (g, f, c). If we are given the values of these parameters, we are given the circle $\mathscr{C}$ represented by the equation $C = 0$. Conversely, the equation of a circle $\mathscr{C}$ can be written in only one way in the form (1). Hence, any circle can be *represented* by the three numbers (g, f, c).

It is natural to think of three numbers (g, f, c) as giving the three coordinates of a point P in three-dimensional space. We shall amplify this statement in a moment. But assuming that we understand what is implied by the term " three-dimensional space ", we can now say that circles in the (x,y)-plane can be *represented* by points in three-dimensional space (x, y, z). In this chapter we shall study the properties of this representation.

2. Euclidean three-space, E_3

Let Ox, Oy, Oz be three mutually perpendicular straight lines drawn through a point O. If P is any point, we consider the

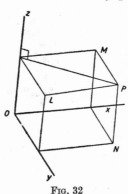

Fig. 32

perpendiculars PL, PM and PN drawn from P to the planes Oyz, Ozx and Oxy respectively. The distance LP measures the

x-coordinate of P, and is positive if LP is in the direction Ox, negative if it is in the opposite direction. Similarly for the y- and z-coordinates of P. If the coordinates of P are (x_1, y_1, z_1), P is the intersection of the three planes $x = x_1$, $y = y_1$, and $z = z_1$.

We do not intend to develop the coordinate geometry of E_3, but a few results are essential for our purpose. The perpendicular from P (x_1, y_1, z_1) on to the line Oz is easily seen to be $\sqrt{(x_1{}^2 + y_1{}^2)}$.

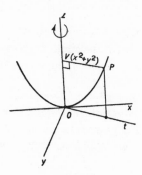

Fig. 33

If we imagine the parabola in the (z, t) plane whose equation is $z - t^2 = 0$, this parabola touches the t-axis at the origin. If we rotate the parabola about the z-axis, we obtain a *paraboloid of revolution*. Since the perpendicular from any point P (x, y, z) of the paraboloid on to the z-axis $= \sqrt{(x^2 + y^2)} = t = \sqrt{z}$, we have
$$x^2 + y^2 - z = 0$$
for the equation of the paraboloid. This surface will play an important part in our investigations.

The only other fact we shall need is that the Joachimstahl ratio-formulae continue to hold in E_3. In other words, if P (x, y, z) and Q (x', y', z') are two points in E_3, the point which divides PQ in the ratio $\lambda : 1$ has coordinates

$$\left(\frac{x + \lambda x'}{1 + \lambda}, \quad \frac{y + \lambda y'}{1 + \lambda}, \quad \frac{z + \lambda z'}{1 + \lambda} \right).$$

This is derived immediately from the formula in the two-dimensional case by projecting P, Q and the point whose coordinates are required on to the plane Oxy, say.

We now return to the representation of a circle $\mathscr{C}$. It is more convenient to take its equation in the form

$$C \equiv x^2 + y^2 - 2\xi x - 2\eta y + \zeta = 0 \qquad (1)$$

rather than that given by Eq. (1), §1. For the centre of $\mathscr{C}$ is (ξ, η), and if we suppose that $\mathscr{C}$ lies in the plane Oxy of E_3, and represent $\mathscr{C}$ by the point P (ξ, η, ζ), *the orthogonal projection of P on to the plane Oxy is the centre of the circle $\mathscr{C}$ which is represented by the point P.*

3. First properties of the representation

The square of the radius of the circle $\mathscr{C}$ given by (1) is $\xi^2 + \eta^2 - \zeta$. Circles of zero radius are represented by points P (x, y, z) which satisfy the equation

$$\Omega \equiv x^2 + y^2 - z = 0. \qquad (1)$$

As we saw in §2, this equation represents a paraboloid of revolution. This paraboloid Ω plays a fundamental part in our investigation. If, at the point $(x, y, 0)$, we erect a perpendicular to the

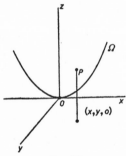

Fig. 34

plane Oxy, this meets Ω in a unique point P, which represents the circle centre $(x, y, 0)$ and of zero radius. Points above P (in an obvious sense) have a z-coordinate which exceeds that of P, and therefore represent circles the square of whose radius is negative. We call such circles *imaginary* circles, noting that the centre of such a circle is a real point. Points below P represent *real* circles, or the ordinary circles of geometry. Since points above P lie inside Ω, and points below P lie outside Ω, we see that *the paraboloid separates the points which represent real circles from those which represent imaginary circles.*

The preceding remarks also show that circles of a system of concentric circles are represented by the points of a line through the common centre and perpendicular to the Oxy-plane.

4. Coaxal systems

The coaxal system defined by the two circles $\mathscr{C}$ and $\mathscr{C}'$, where $\mathscr{C}'$ is given by the equation

$$C' \equiv x^2 + y^2 - 2\xi'x - 2\eta'y + \zeta' = 0,$$

is given by the equation

$$x^2 + y^2 - 2\xi x - 2\eta y + \zeta + \lambda[x^2 + y^2 - 2\xi'x - 2\eta'y + \zeta'] = 0,$$

which, for our purpose, must be written in the form

$$x^2 + y^2 - 2\left[\frac{\xi + \lambda\xi'}{1 + \lambda}\right]x - 2\left[\frac{\eta + \lambda\eta'}{1 + \lambda}\right]y + \left[\frac{\zeta + \lambda\zeta'}{1 + \lambda}\right] = 0.$$

The representative point is therefore

$$\left(\frac{\xi + \lambda\xi'}{1 + \lambda}, \ \frac{\eta + \lambda\eta'}{1 + \lambda}, \ \frac{\zeta + \lambda\zeta'}{1 + \lambda}\right).$$

This point is on the line PP', where P represents $\mathscr{C}$, and P' represents $\mathscr{C}'$. Hence *the circles of a coaxal system are represented by the points of a line in E_3.* Any two distinct points of a given line in E_3 determine the line. Hence we see once more that any two distinct circles of a coaxal system determine the coaxal system.

The line PP' will meet the paraboloid Ω in two, one or no real points. Intersections with Ω correspond to zero circles, or limiting points, in the coaxal system. Hence the line representing a non-intersecting system of coaxal circles meets Ω in real

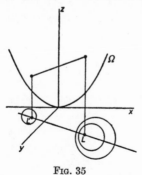

Fig. 35

points, whereas the line representing an intersecting system of coaxal circles does not meet Ω.

A pair of points in the Oxy-plane determines a coaxal system by the condition that the circles of the system should pass through both points. This coaxal system determines a line in E_3. Two lines in E_3 which meet represent coaxal systems which have one circle in common. If both systems are of intersecting types, and the common points of one are P, P', and of the other Q, Q', it follows in this case that P, P', Q, Q' are concyclic.

Bearing these facts in mind, we see that known theorems in E_3 yield theorems for coaxal systems of circles. We give two examples.

5. Deductions from the representation

Let l, m be two skew lines in E_3, and V a point which does not lie on either line. Through V we can draw a unique line n which

Fig. 36

will meet both l and m, that is, a unique *transversal* of l and m. This line n must lie in the plane obtained by joining l to V, and it must also lie in the plane obtained by joining m to V. These two planes intersect in a line through V, and this is the required line n. If there were two such lines n, l and m would be coplanar, contrary to hypothesis.

We translate this theorem into its plane equivalent:

Let l and m both represent intersecting coaxal systems, one defined by the points P, P', the other by the points Q, Q'. Since l and m do not meet, the coaxal systems do not have a common circle, and the four points P, P', Q, Q', are not concyclic. The point V represents a circle $\mathscr{C}$ which is not a member of either coaxal system. The line n gives a coaxal system which contains $\mathscr{C}$. If this third system is also of the intersecting type, let R, R' be the points which define it. These points lie on $\mathscr{C}$.

Since n meets both l and m, there is a circle common to the systems represented by l and n, and there is also a circle common to the systems represented by m and n.   Hence P, P', R, R' are concyclic, and so are Q, Q', R, R'.   The plane theorem becomes:

P, P', Q, Q' *are four points which are not concyclic, and do not lie on a given circle* $\mathscr{C}$.   *Then there is a unique pair of points* R, R' *on* $\mathscr{C}$ *such that* P, P', R, R' *and* Q, Q', R, R' *are respectively concyclic.*

This theorem may, of course, be proved directly.   If we draw circles through P, P', their common chords with $\mathscr{C}$ form a pencil of lines.   This hint should enable the reader to complete the proof.

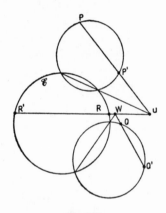

Fig. 37

Our second example is derived from a theorem in E_3 which it would take us too long to prove here:

Four lines in E_3 *have two, one, or an infinity of transversals.*

The proof involves the general theory of quadric surfaces, and is beyond our scope in this book.   From the preceding discussion we rapidly arrive at the following theorem:

Given four pairs of points P, P', Q, Q', R, R', S, S' *in a plane, the problem of finding another pair* X, X' *such that* $PP'XX'$, $QQ'XX'$, $RR'XX'$ *and* $SS'XX'$ *are respectively concyclic admits of two, one or an infinity of solutions.*

It would not be so easy to give a direct proof of this theorem.

6. Conjugacy relations

Let $\mathscr{C}_1$ and $\mathscr{C}_2$ be two circles. Then, with the usual notation,

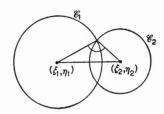

Fig. 38

we find the condition that they intersect at an angle α. The respective centres are (ξ_1, η_1) and (ξ_2, η_2), and the radii from the centres to a point of intersection enclose an angle α or $\pi - \alpha$. The cosine formula gives the relation

$$(\xi_1 - \xi_2)^2 + (\eta_1 - \eta_2)^2 = (\xi_1{}^2 + \eta_1{}^2 - \zeta_1) + (\xi_2{}^2 + \eta_2{}^2 - \zeta_2)$$
$$\pm\, 2(\xi_1{}^2 + \eta_1{}^2 - \zeta_1)^{\frac{1}{2}}\, (\xi_2{}^2 + \eta_2{}^2 - \zeta_2)^{\frac{1}{2}} \cos\alpha.$$

This becomes

$$2\xi_1\xi_2 + 2\eta_1\eta_2 - \zeta_1 - \zeta_2$$
$$= \mp\, 2(\xi_1{}^2 + \eta_1{}^2 - \zeta_1)^{\frac{1}{2}}\, (\xi_2{}^2 + \eta_2{}^2 - \zeta_2)^{\frac{1}{2}} \cos\alpha,$$

which is more useful in the form

$$(2\xi_1\xi_2 + 2\eta_1\eta_2 - \zeta_1 - \zeta_2)^2$$
$$= 4(\xi_1{}^2 + \eta_1{}^2 - \zeta_1)\,(\xi_2{}^2 + \eta_2{}^2 - \zeta_2)\cos^2\alpha. \qquad (1)$$

We first examine two special cases:

(i) $\alpha = 0$. The circles $\mathscr{C}_1$ and $\mathscr{C}_2$ touch. If we take the variable point $\left(\dfrac{\xi_1 + \lambda\xi_2}{1 + \lambda},\ \dfrac{\eta_1 + \lambda\eta_2}{1 + \lambda},\ \dfrac{\zeta_1 + \lambda\zeta_2}{1 + \lambda}\right)$ on the line P_1P_2, this point lies on Ω if and only if

$$(\xi_1 + \lambda\xi_2)^2 + (\eta_1 + \lambda\eta_2)^2 - (1 + \lambda)\,(\zeta_1 + \lambda\zeta_2) = 0.$$

As a quadratic in λ this becomes

$$\lambda^2(\xi_2{}^2 + \eta_2{}^2 - \zeta_2) + \lambda(2\xi_1\xi_2 + 2\eta_1\eta_2 - \zeta_1 - \zeta_2)$$
$$+ (\xi_1{}^2 + \eta_1{}^2 - \zeta_1) = 0. \qquad (2)$$

The condition (1) with $\cos\alpha = 1$ is just the condition that the quadratic (2) should have equal roots. Hence, if the circles $\mathscr{C}_1$ and $\mathscr{C}_2$ touch, the line P_1P_2 touches Ω. But the intersections of this line with Ω give the limiting points of the coaxal system determined by $\mathscr{C}_1$ and $\mathscr{C}_2$. We deduce that a necessary and sufficient condition that two circles touch is that the limiting points of the coaxal system they determine should coincide.

(ii) $\alpha = \frac{1}{2}\pi$. In this case (1) becomes simply
$$2\xi_1\xi_2 + 2\eta_1\eta_2 - \zeta_1 - \zeta_2 = 0. \tag{3}$$
From (2) we see that this condition implies that the roots of the quadratic are of the form $k, -k$. Hence the intersections of P_1P_2 with Ω divide P_1P_2 internally and externally in the same ratio. That is, the four points form a harmonic range, and P_1, P_2 are *conjugate* points with respect to Ω.

Hence *orthogonality implies conjugacy, and conversely.*

The locus of points conjugate, with respect to Ω, to the fixed point (ξ',η',ζ') is given by the equation
$$2x\xi' + 2y\eta' - z - \zeta' = 0. \tag{4}$$
This represents a plane in E_3, the *polar* plane of (ξ',η',ζ'). The points of this plane represent the circles orthogonal to the circle
$$x^2 + y^2 - 2\xi'x - 2\eta'y + \zeta' = 0. \tag{5}$$
The plane given by (4) meets Ω in a conic. If (x,y,z) is a point on this conic, $z = x^2 + y^2$, and since (4) is satisfied, we have
$$2x\xi' + 2y\eta' - (x^2 + y^2) - \zeta' = 0,$$
which is the same equation as (5). The projection of the point (x,y,z) on to the plane Oxy is the point $(x,y,0)$, and we have proved

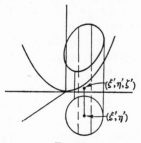

Fig. 39

that *the projection of the conic in which the polar plane of a point in E_3 meets Ω is precisely the circle represented by the point.*

We may see this from another point of view. The points of the conic cut on Ω represent *point* circles, that is, circles of zero radius, orthogonal to the circle given by (5). It is easily verified that if a point circle is orthogonal to a given circle, the centre of the point circle must lie *on* the given circle. Hence the projection of the conic on to the plane Oxy is bound to give the circle itself.

The polar plane of the point
$$\left(\frac{\xi_1 + \lambda\xi_2}{1 + \lambda}, \quad \frac{\eta_1 + \lambda\eta_2}{1 + \lambda}, \quad \frac{\zeta_1 + \lambda\zeta_2}{1 + \lambda}\right),$$
which, as λ varies, describes the line P_1P_2, is
$$2\xi\left(\frac{\xi_1 + \lambda\xi_2}{1 + \lambda}\right) + 2\eta\left(\frac{\eta_1 + \lambda\eta_2}{1 + \lambda}\right) - \zeta - \left(\frac{\zeta_1 + \lambda\zeta_2}{1 + \lambda}\right) = 0.$$
This may be written in the form
$$2\xi\xi_1 + 2\eta\eta_1 - \zeta - \zeta_1 + \lambda(2\xi\xi_2 + 2\eta\eta_2 - \zeta - \zeta_2) = 0,$$
which shows that the polar plane always passes through a fixed line, given by the intersection of the polar planes of P_1 and of P_2. If we call this line l', and call P_1P_2 l, any point on l is conjugate to every point of l', and therefore any point on l' is conjugate to every point of l. Two such lines are called *polar* lines, and they represent conjugate coaxal systems. If l meets Ω in the points P, P', the polar planes of these points are the respective tangent planes at the points to Ω, and these tangent planes will contain l'. If l meets Ω, and so represents a non-intersecting system of coaxal circles, l' will not meet Ω.

The tangent plane at P represents the circles orthogonal to the point-circle represented by P, that is, the circles which *pass through* the projection of P on to the Oxy-plane. Hence, l' represents the circles which pass through the limiting points of the coaxal system represented by l.

Again, the projection of l on to the plane Oxy gives the line of centres of the coaxal system represented by l. Hence *polar lines project into perpendicular lines.*

We can easily find a tetrahedron in E_3 which is self-polar: that is, every face is the polar plane of the opposite vertex. Choose any point P not on Ω, and let π be its polar plane. Choose any point Q in π not on Ω, and let π' be its polar plane. π' contains P and meets π in a line l. Choose any point R on l not on Ω. Its polar plane contains P and Q and meets l in S. Then it is clear that $PQRS$ is a self-polar tetrahedron.

Since polar lines project into perpendicular lines, and the opposite edges of $PQRS$ are polar lines, we see that *if four circles are mutually orthogonal, each being orthogonal to the other three, then their centres form a triangle and its orthocentre.* It can be shown that one vertex of a self-polar tetrahedron must lie *inside* Ω. Hence, one of the four mutually orthogonal circles must be imaginary.

7. Circles cutting at a given angle

We now return to the general relation (1) of the previous section, and see that the points representing circles which cut a fixed circle $\mathscr{C}'$ at an angle α or $\pi - \alpha$ satisfy the equation

$$4(x^2 + y^2 - s)(\xi'^2 + \eta'^2 - \zeta')\cos^2\alpha$$
$$= (2x\xi' + 2y\eta' - s - \zeta')^2. \qquad (1)$$

This represents a quadric surface, and the form of its equation shows that it has *ring-contact* with Ω along the intersection of Ω with the plane

$$2x\xi' + 2y\eta' - s - \zeta' = 0. \qquad (2)$$

If the reader is unfamiliar with the idea of ring-contact he need not be disturbed, since it is only the algebraic form of (1) which will concern us. We know already that the curve in which the plane given by (2) meets Ω projects into the circle $\mathscr{C}'$ itself.

We wish to investigate the circles which cut three given circles $\mathscr{C}_1$, $\mathscr{C}_2$ and $\mathscr{C}_3$ at angles α or $\pi - \alpha$. To do this, we must examine the common intersections of the three quadric surfaces:

$$4(x^2 + y^2 - s)(\xi_r^2 + \eta_r^2 - \zeta_r)\cos^2\alpha$$
$$= (2x\xi_r + 2y\eta_r - s - \zeta_r)^2 \qquad (r = 1, 2, 3).$$

For the sake of brevity we write these equations as

$$4(x^2 + y^2 - s)k_r\cos^2\alpha = X_r^2 \qquad (r = 1, 2, 3), \qquad (3)$$

where k_r is a constant, and $X_r = 0$ is the polar plane of (ξ_r, η_r, ζ_r). If we equate the common value of $x^2 + y^2 - s$ in the three equations, we obtain the equations

$$\frac{X_1^2}{k_1} = \frac{X_2^2}{k_2} = \frac{X_3^2}{k_3}.$$

We note that these equations do not depend on $\cos\alpha$. If we take them in pairs, the system is equivalent to

$$\begin{cases} \sqrt{k_2}\,X_1 \pm \sqrt{k_1}\,X_2 = 0, \\ \sqrt{k_3}\,X_2 \pm \sqrt{k_2}\,X_3 = 0. \end{cases} \qquad (4)$$

Taking all the possible combinations of signs, we have four sets of two linear equations in x, y, s. Since two planes determine a line, the solutions of (4) lie on four straight lines. Since $X_1 = X_2 = X_3 = 0$ automatically satisfies (4), these four lines all pass through the point determined by the intersection of the polar planes of (ξ_r, η_r, ζ_r) $(r = 1, 2, 3)$. This point represents the circle orthogonal to the three given circles.

Each one of the lines determined by (4) meets each one of the three quadrics given by (3) in the same two points. Hence there are *eight* solutions of the equations, for a given value of cos α, and we have proved the following theorem:

There are eight circles which intersect three given circles at a given angle α or π — α. As the angle α varies the circles vary in four coaxal systems with a common circle, the circle orthogonal to the three given circles.

The case α = 0 gives the solution of the problem of Apollonius (see chapter I, §10). Since each one of the lines determined by (4) gives two solutions of (3), each one of the four coaxal systems contains two of the circles which satisfy the given conditions.

8. Representation of inversion

Let A, A' be inverse points in a circle $\mathscr{C}$. Then $\mathscr{C}$ is orthogonal to any circle through A and A', and $\mathscr{C}$ is a member of the coaxal ·system which is determined by the limiting points A, A'. Let P represent the circle $\mathscr{C}$. The zero circles centres A, A' are represented by points on Ω. By the remark above, P is collinear with these two points on Ω.

Since P is regarded as fixed, and A, A' as variable, we have the following method for finding the inverse, in a given circle $\mathscr{C}$, of a given curve, $\mathscr{D}$, say:

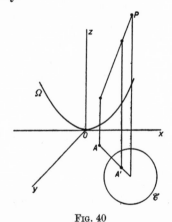

Fig. 40

Erect perpendiculars to the plane Oxy at the points of $\mathscr{D}$, and let $\mathscr{E}$ be the curve in which these perpendiculars meet Ω. The cone of lines joining P to the points of $\mathscr{E}$ meets Ω in a further curve $\mathscr{E}'$. Project $\mathscr{E}'$ down on to the plane Oxy. The resulting curve $\mathscr{D}'$ is the inverse of $\mathscr{D}$ in $\mathscr{C}$.

We remark that this construction shows at once that the eight circles which were discussed in the preceding section are inverse in pairs in the common orthogonal circle of the three circles. Since each of the three given circles inverts into itself with respect to the common orthogonal circle, this result is to be expected.

If the curve $\mathscr{D}$ to be inverted is a circle, we know that the curve $\mathscr{E}$ obtained by projecting it on to Ω is a plane section. The cone of lines joining P to the points of $\mathscr{E}$ is a quadric cone, and since the inverse of a circle is another circle, this cone meets Ω again in a further plane section $\mathscr{E}'$. Again, a circle, the circle of inversion and the inverse circle are coaxal, so that we obtain the theorem:

A quadric cone which meets Ω in one plane section will meet it again in another plane section, and the poles of the two plane sections are collinear with the vertex of the cone.

The more advanced reader will realize that this is a *projective* theorem, and is therefore true for any quadric Ω.

9. The envelope of a system

We may measure the complexity of a system of circles by the complexity of the curve in E_3 which represents the system. From this point of view a coaxal system of circles is the simplest kind of system, since a line is the simplest curve in E_3. A coaxal system of circles has no *envelope*: that is, there is no curve touched by all circles of the system. For intersecting systems we may, if we wish, say that the envelope consists of the two points common to all circles of the system. But interesting envelopes only appear when we consider *conic systems of circles*.

A conic system of circles is one represented by a conic in E_3. A conic is a plane curve, and this plane will have a pole with respect to Ω. This pole represents a circle which is orthogonal to all the circles of the conic system. Again, the projection of the conic on to the plane Oxy is, except in one special case, a conic, and gives the locus of centres of circles of the conic system.

Hence the centres of circles of a conic system lie on a conic, and all circles of the system are orthogonal to a fixed circle.

These two properties serve to define a conic system, since the second property restricts the representative points to lie in a plane, and the intersection of a conical cylinder and a plane is a conic.

We now find the envelope of a conic system of circles. Let the curve touched by all circles of the system be $\mathscr{E}$, and suppose that the circle $\mathscr{C}$ touches it at the point A. A circle of the system

FIG. 41

" near " to $\mathscr{C}$ will intersect $\mathscr{C}$ in a point which is " near " to A. In the limit the intersection of $\mathscr{C}$ and a neighbouring circle of the conic system includes the point of contact of $\mathscr{C}$ with $\mathscr{E}$. It can be shown that all circles of the system touch the locus of ultimate intersections of neighbouring circles, so that all circles of the system touch the envelope at *two* points.

Let C be the conic in E_3 which represents the conic system of circles. If P is the point on C which represents $\mathscr{C}$, the tangent to C at P represents the coaxal system defined by $\mathscr{C}$ and its neighbouring circle. The common points of this coaxal system are the points on $\mathscr{E}$ we wish to determine. Now the line in E_3 polar to the tangent to C at P (with respect to Ω) represents the *conjugate* coaxal system, whose limiting points are the points on $\mathscr{E}$ required. If we find the intersections of this polar line with Ω, and project down on to the plane Oxy, we have the points we want.

Therefore, *to obtain the envelope of the system of circles represented by a curve* C *in* E_3, *we find the curve in which lines polar to the tangents of* C *(with respect to* Ω*) meet* Ω, *and project this curve orthogonally on to the* (x,y)-*plane.*

The method applies to any curve C, but we are interested in the case in which C is a conic. Then the lines polar to the tangents of C all pass through a fixed point, the pole of the plane of C, and generate a quadric cone. Hence, the envelope is the orthogonal projection of the curve in which a quadric cone meets Ω.

Now two quadrics intersect in a curve which is cut by a plane in four points, these points being the four intersections of the respective conics in which the plane cuts the quadrics. Such a curve is called a *quartic* curve. If the quadric intersecting Ω has the equation

$$Q \equiv ax^2 + by^2 + cz^2 + 2fyz + 2gzx + 2hxy + 2ux$$
$$+ 2vy + 2wz + d = 0,$$

the orthogonal projection of the intersection with Ω is obtained by replacing z, wherever it occurs in the equation, by $x^2 + y^2$. We therefore obtain the curve

$$ax^2 + by^2 + c(x^2 + y^2)^2 + 2fy(x^2 + y^2) + 2g(x^2 + y^2)x$$
$$+ 2hxy + 2ux + 2vy + 2w(x^2 + y^2) + d = 0.$$

This curve has double points at the circular points at infinity, and is called a *bicircular quartic curve*. The envelope of a conic system of circles is therefore a bicircular quartic curve.

The quartic curve on Ω is the intersection of two quadrics $\Omega = 0$ and $Q = 0$, and therefore the pencil of quadrics $\Omega + \lambda Q = 0$ passes through it. A pencil of quadrics contains four cones. Hence four cones intersect Ω in the quartic curve whose projection gives the envelope. One of these cones has appeared already, its generators being the polar lines of the tangents to the conic C which specified the conic system of circles. Since each quadric cone can be derived as the set of polar lines of tangents to a conic, we see that the bicircular quartic curve may be regarded as the envelope of *four* systems of circles.

Again, a conic in E_3 meets Ω in four points. Hence a conic system of circles contains four point-circles. Since a point-circle

$$(x - a)^2 + (y - b)^2 = 0$$

may also be written in the form $(i = \sqrt{-1})$:

$$[(x - a) + i(y - b)][(x - a) - i(y - b)] = 0,$$

a point-circle may also be regarded as the isotropic lines through the centre (a,b). If these touch a given curve, the point (a,b) is, by definition, a focus of the curve.

Hence the bicircular quartic we have obtained has sixteen foci.

It is easy to see that any bicircular quartic may be obtained in this way, and therefore has the properties described above, but we shall not pursue the matter any further.

10. Some further applications

Let us suppose that we are given a plane algebraic curve, of equation $f(x,y) = 0$. This curve is touched by an infinity of circles through the point $(0,0)$. If such a circle is

$$x^2 + y^2 - 2\xi x - 2\eta y = 0,$$

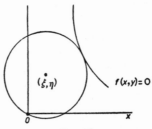

FIG. 42

a relation $g(\xi,\eta) = 0$ holds; that is, *the centres of circles through* $(0,0)$ *which touch a given curve* $f(x,y) = 0$ *lie on a curve* $g(x,y) = 0$.

We may call this latter curve *the circle-tangential equation* of the given curve. We consider some examples.

If the curve $f(x,y) = 0$ consists merely of the point (a,b), the circle through $(0,0)$ has to pass through (a,b). The locus of centres

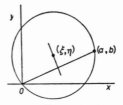

FIG. 43

of such circles is the perpendicular bisector of the join of (a,b) and $(0,0)$.

If $f(x,y) = 0$ represents a line, the circle-tangential equation is

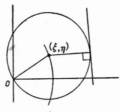

FIG. 44

the locus of a point which moves so that its distance from a fixed point is always equal to its distance from a fixed line. Thus the locus is a parabola, with (0,0) as focus and the line as directrix.

Finally, we consider the case when $f(x,y) = 0$ is the circle

$$x^2 + y^2 - 2\xi'x - 2\eta'y + \zeta' = 0.$$

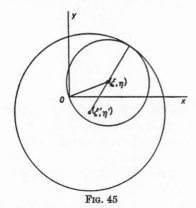

FIG. 45

There are two cases:

(i) If (0,0) lies inside the circle ($\zeta' < 0$), the circle-tangential equation is the locus of a point which moves so that the *sum* of its distances from two fixed points, (0,0) and (ξ',η'), is constant. The locus is therefore an ellipse with these points as foci.

(ii) If (0,0) lies outside the circle ($\zeta' > 0$), it is the *difference* of the distances from (0,0) and (ξ',η') which is constant. The circle-tangential equation is then a hyperbola with these points as foci.

We may apply the methods of the preceding section to finding the circle-tangential equation of a given curve $f(x,y) = 0$. When applied to the above examples, we shall obtain some properties of the paraboloid Ω.

Let us suppose then that $g(x,y) = 0$ is the circle-tangential equation of $f(x,y) = 0$. Since all the circles we are considering pass through (0,0), *the curve $g(x,y) = 0$ is itself the representative curve in E_3 of a system of circles which have $f(x,y) = 0$ as envelope.* Applying the method of the last section, we find the polar lines of the tangents to $g(x,y) = 0$. We obtain a cone of lines, vertex (0,0). This cone meets Ω in a curve of which $f(x,y) = 0$, the envelope, is the orthogonal projection on to the (x,y)-plane.

Since we are given $f(x,y) = 0$, to obtain $g(x,y) = 0$ we project

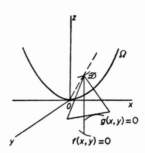

FIG. 46

the given curve up on to Ω, obtaining a curve $\mathscr{D}$, say. The tangent planes to Ω at the points of $\mathscr{D}$ meet the (x,y)-plane in the tangents of the required circle-tangential equation. This follows from the fact that the polar line of a line passing through $(0,0)$ is the intersection of the polar planes of $(0,0)$ and the point where this line meets Ω again. The polar plane of $(0,0)$ is the (x,y)-plane, and the polar plane of any other point on Ω is also the tangent plane at the point.

We apply this method to the examples given above. For the circle-tangential equation of a point (a,b), we consider the intersection with $z = 0$ of the tangent plane to Ω at the point $(a,b, a^2 + b^2)$. This intersection must be the perpendicular bisector of the join of (a,b) to $(0,0)$.

We now operate with the circle $x^2 + y^2 - 2\xi'x - 2\eta'y + \zeta' = 0$. When we project this up on to Ω, we obtain a plane section given by the intersection of Ω and the plane
$$2\xi'x + 2\eta'y - z - \zeta' = 0.$$
Tangent planes to Ω at the points of this section all pass through the pole of the plane, which is the point (ξ',η',ζ'), and meet $z = 0$ in tangents to the curve given by the intersection with $z = 0$ of the tangent cone to Ω from (ξ',η',ζ').

The tangent cone is easily derived from Eq. (2), §6, chapter II, and is given by the equation
$$4(\xi'^2 + \eta'^2 - \zeta')(x^2 + y^2 - z) - (2\xi'x + 2\eta'y - \zeta' - z)^2 = 0.$$
If we put $z = 0$ in this equation, we obtain
$$4(\xi'^2 + \eta'^2 - \zeta')(x^2 + y^2) - (2\xi'x + 2\eta'y - \zeta')^2 = 0.$$
Our previous investigation tells us that this is *a conic with foci at* $(0,0)$ *and* (ξ',η'), *and an ellipse or a hyperbola according as* ζ' *is negative or positive.*

The presence of a focus at $(0,0)$ is manifest, since the equation may be written as

$$x^2 + y^2 = \frac{(\xi'^2 + \eta'^2)}{(\xi'^2 + \eta'^2 - \zeta')} \left[\frac{2\xi'x + 2\eta'y - \zeta'}{2\sqrt{(\xi'^2 + \eta'^2)}} \right]^2.$$

The corresponding directrix is the intersection with $s = 0$ of the polar plane of (ξ',η',ζ'). We have found then that the tangent cone to Ω from the point (ξ',η',ζ') meets $s = 0$ in a conic with foci at $(0,0)$ and (ξ',η'), and that this conic is an ellipse or hyperbola according as ζ' is negative or positive.

11. Some anallagmatic curves

We conclude this chapter by finding the equations of curves of a given order which are anallagmatic (self-inverse) in a given circle.

Let the circle have representative point (ξ,η,ζ). As we saw in §8, page 36, points which are inverse in the circle

$$x^2 + y^2 - 2\xi x - 2\eta y + \zeta = 0$$

are represented by points on Ω collinear with the point (ξ,η,ζ).

A plane through (ξ,η,ζ) is given by the equation

$$p(x - \xi) + q(y - \eta) + r(s - \zeta) = 0.$$

This plane intersects Ω in a conic, the orthogonal projection of which is the circle

$$p(x - \xi) + q(y - \eta) + r(x^2 + y^2 - \zeta) = 0.$$

By construction, this circle is self-inverse, and therefore orthogonal to the given circle. It is the most general circle orthogonal to the given circle.

A quadric cone with vertex at (ξ,η,ζ) has the equation

$$a(x - \xi)^2 + 2h(x - \xi)(y - \eta) + b(y - \eta)^2$$
$$+ 2g(x - \xi)(s - \zeta) + 2f(y - \eta)(s - \zeta) + c(s - \zeta)^2 = 0,$$

and therefore the quartic curves anallagmatic in the given circle are obtained by merely substituting $s = x^2 + y^2$ in this equation. They are bicircular quartics.

To obtain the anallagmatic *cubic* curves, the quartic curve of intersection of the quadric cone vertex (ξ,η,ζ) with Ω must have one point at infinity on the axis, $x = y = 0$, of Ω. Therefore, one generator of the cone must be parallel to $x = y = 0$. We therefore put $c = 0$ in the above equation to obtain anallagmatic cubics.

This process can easily be extended.

CHAPTER III

Our aim in this chapter is to discuss the Poincaré model of hyperbolic non-Euclidean geometry. This sounds more formidable than it really is. Circles play a leading part in the discussion, as we shall see, and so do the elementary properties of complex numbers. We begin by recapitulating these properties.

1. Complex numbers

Complex numbers are expressions of the form $a + bi$, where a and b are any real numbers, the rules of calculation for these complex numbers being the same as for real numbers, plus the rule that $i \cdot i = i^2 = -1$.

Hence, we have
$$(a + bi) + (c + di) = (a + c) + (bi + di) = (a + c) + (b + d)i,$$
and
$$(a + bi)(c + di) = ac + bci + adi + bdi^2$$
$$= ac + bci + adi - bd$$
$$= (ac - bd) + (bc + ad)i.$$

We say that a is the *real part*, bi the *imaginary part* of the complex number $a + bi$. A complex number is only zero if its real part and imaginary part are simultaneously zero. Hence, $a + bi = c + di$ if and only if $a = c$ and $b = d$.

2. The Argand diagram

If, in a plane, a system of rectangular Cartesian axes be taken, the complex number $z = x + iy$ may be represented by the point with coordinates (x, y). In this way we set up a correspondence between the complex numbers and points of the plane. In particular, the numbers $x + i \cdot 0$ correspond to points of Ox, which is therefore called the *real axis*. Such numbers are indistinguishable from ordinary real numbers x.

This method of representing complex numbers is called the Argand diagram. The plane thus put into correspondence with the set of complex numbers is called *the plane of the complex variable*, and we shall speak of *the complex number z* or of *the point*

44

z, this point z being precisely the point which corresponds to the number z.

It is sometimes useful to think of the point z as a *vector $\overrightarrow{Oz}$* with components (x, y). We recall that two vectors are equal if they have the same components, that is to say if the arrows which represent them are parallel, in the same sense, and of equal length, without necessarily starting from the same origin.

We see at once that these vectors obey the parallelogram law

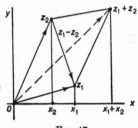

FIG. 47

of addition, since if $z_1 = x_1 + iy_1$, and $z_2 = x_2 + iy_2$,
$$z = z_1 + z_2 = x_1 + x_2 + i(y_1 + y_2).$$
The interpretation of subtraction is equally simple:

If $z = z_1 - z_2$, and therefore $z_1 = z + z_2$, z must be represented by the vector $\overrightarrow{z_2 z_1}$.

3. Modulus and argument

If the length of the vector $\overrightarrow{Oz}$ be denoted by ρ, and the angle

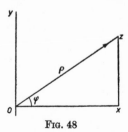

FIG. 48

through which Ox must be rotated to lie along $\overrightarrow{Oz}$ by φ, we have
$$x = \rho \cos \varphi, \qquad y = \rho \sin \varphi, \qquad \rho = \sqrt{(x^2 + y^2)}.$$
ρ is called the *modulus* of the complex number, and φ, which is determined to a multiple of 2π, the *argument*.

It follows that
$$s = x + iy = \rho(\cos \varphi + i \sin \varphi).$$
If now s_1 and s_2 are two complex numbers,
$$s_1 = \rho_1(\cos \varphi_1 + i \sin \varphi_1),$$
$$s_2 = \rho_2(\cos \varphi_2 + i \sin \varphi_2),$$
their product $s_1 s_2 = \rho_1 \rho_2 (\cos (\varphi_1 + \varphi_2) + i \sin (\varphi_1 + \varphi_2))$, as is easily seen by direct multiplication. Hence, to multiply two complex numbers, you multiply their moduli and add their arguments.

For division, let $s = s_1/s_2$, $(s_2 \neq 0)$, so that
$$s_1 = s s_2.$$
Using the first rule, it follows that the quotient of two complex numbers has, for modulus, the quotient of their moduli, and, for argument, the difference of their arguments.

4. Circles as level curves

Let a, b, s be any three complex numbers. We denote by (sab) the ratio

$$\frac{s - a}{s - b} = (sab).$$

The modulus of this complex number is the quotient of the moduli of $s - a$ and $s - b$. It is therefore the quotient

$$as^\dagger / bs \, .$$

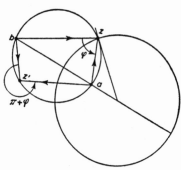

Fig. 49

---

† as denotes the distance between the points a and s in the Argand diagram.

As for its argument, by the above it is the angle through which the vector $z - b$ must be turned to give it the direction of the vector $z - a$. It is therefore the angle between the vectors $\overrightarrow{az}$ and $\overrightarrow{bz}$, oriented as in Fig. 49.

If we regard a and b as fixed, and z as variable, we may enquire into the nature of the curves along which mod (zab) and arg (zab) are respectively constant. These are the *level curves* of the respective functions. Both turn out to be circles.

If we have $az/bz = $ constant, the ratio of the powers of the point z with respect to the point circles a and b is constant. Therefore z moves on a circle of the coaxal system determined by the point circles a,b. This coaxal system has these point circles as limiting points.

If the angle between $\overrightarrow{az}$ and $\overrightarrow{bz}$ is constant, z moves on an arc of a certain circle passing through a and b. The other arc, the arc complementary to the first, corresponds to an argument $\varphi + \pi$. If φ takes all possible values, we obtain all circles of the coaxal system through a and b, including the radical axis ab, when $\varphi = \pi$ (this gives points between a and b) and when $\varphi = 0$ (this gives points outside ab).

For all values of the modulus and argument of (zab) we therefore obtain all circles of two conjugate coaxal systems as level curves.

5. The cross-ratio of four complex numbers

Let a, b, c, d be four complex numbers. By definition, the cross-ratio of the four numbers in the given order is the ratio $(acd)/(bcd)$. We write

$$\frac{(acd)}{(bcd)} = (ab,cd),$$

so that

$$(ab,cd) = \frac{(a-c)/(a-d)}{(b-c)/(b-d)}.$$

This is formally the same definition as that which occurs in projective geometry.

The *modulus* of the cross-ratio is equal to the quotient of the moduli of (acd) and (bcd). Hence, for this modulus to be equal to

1, it is necessary and sufficient for (acd) and (bcd) to have the same modulus. By what we have seen above, a and b must therefore lie on the same circle of the coaxal system which has c,d as limiting points.

The *argument* of the cross-ratio is equal to the difference of the arguments of (acd) and (bcd). The cross-ratio therefore has zero argument if (acd) and (bcd) have the same argument, which is the case if a and b are on the same arc of a circle bounded by c and d.

The argument of the cross-ratio is equal to π if the arguments

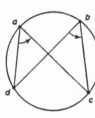 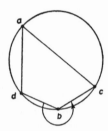

FIG. 50   FIG. 51

of (acd) and (bcd) differ by π, which is the case if a and b are on complementary arcs of a circle through c and d.

Now, a number is real if its argument is 0 or π. Hence we have the theorem:

For the cross-ratio (ab,cd) to be real, it is necessary and sufficient that the four points a, b, c, d should lie on a circle (or straight line).

More generally, we find the locus of z such that the cross-ratio (za,bc) has (i) a constant modulus; and (ii) a constant argument.

Since $(za,bc) = (zbc)/(abc)$, the modulus and argument of (za,bc) will be constant if the respective modulus and argument of (zbc) are constant.

(i) The locus of points z such that mod (zbc) is constant is a circle of the coaxal system with limiting points b and c.

(ii) The locus of points z such that arg (zbc) is constant is an arc of a circle limited by b and c.

We deduce that if a, b, c are three distinct complex numbers, u any given complex number, then *there is a unique number z such that*

$$(za,bc) = u.$$

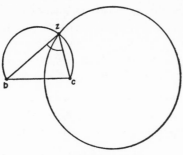

FIG. 52

For u has a definite modulus and argument, and z is given as the unique intersection of a circle and an arc of a circle, as shown in Fig. 52.

This important property is worth verifying algebraically. If $(za,bc) = u$, then we must have $(zbc) = (abc)u$. Write $(abc)u = k$. Then $(z - b)/(z - c) = k$, which gives

$$z = \frac{b - kc}{1 - k}.$$

We conclude this section by giving an interesting geometrical

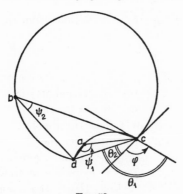

FIG. 53

interpretation to the cross-ratio (ab,cd) of any four complex numbers. If ψ_1 is the argument of (acd), and ψ_2 that of (bcd), we know that the argument φ of (ab,cd) satisfies the equation

$$\varphi = \psi_1 - \psi_2.$$

Draw the circles through a, c, d and b, c, d respectively, and let Θ_1, Θ_2 be the angles made by cd with the tangents at c to these

circles. Then by the alternate segment theorem, $\psi_1 = \Theta_1$ and
$\psi_2 = \Theta_2$, so that $\varphi = \Theta_1 - \Theta_2$. Hence φ is the angle between
the two circles, and we have the theorem:

*The argument of the cross-ratio (ab,cd) is equal to the angle
between the two circles passing through a, c, d and b, c, d respectively.*

6. Möbius transformations of the z-plane

These are transformations of the plane of the complex variable
which map any point z on a point w according to the formula

$$w = \frac{Az + B}{Cz + D}, \tag{1}$$

where A, B, C, D are four complex constants, and $AD - BC \neq 0$.

We exclude the case $A/C = B/D$, because then $w = A/C$ for all
values of z, and all points z transform into a single point w.

The reader is probably already familiar with transformations
of type (1) in projective geometry, where they are called *bilinear*
or *projective* transformations, and transform points of a line. The
same algebra proves the theorem:

A Möbius transformation leaves cross-ratios invariant.

In other words, if a, b, c, d transform respectively into $a', b', c',$
d', then $(ab,cd) = (a'b',c'd')$.

We make some important deductions from this theorem.

I. If $w = \dfrac{Az + B}{Cz + D}$, $(AD - BC \neq 0)$ is a Möbius transforma-

tion, and $(a,a'), (b,b'), (c,c')$ are three pairs of corresponding points,
then the equation of the transformation may be written in the form

$$(wa',b'c') = (za,bc). \tag{2}$$

This follows immediately from the theorem. If z is any point,
and w the corresponding point under the Möbius transformation,
then the pair (z,w) satisfy (2). On the other hand we know that,
z having been given, there is a unique w which can make the
cross-ratio $(wa',b'c') = (za,bc)$. Hence the equation (2) does, in
fact, map z on the w given by the Möbius transformation.

II. Conversely, let a, b, c be three distinct complex numbers,
and a', b', c' a second set of three distinct complex numbers.
Then the equation

$$(wa',b'c') = (za,bc)$$

is the equation of a Möbius transformation which maps a on a',
b on b' and c on c'.

This follows immediately by writing out both sides of the equation, and substituting $z = a, b, c$ in turn. We find that $w = a'$, b', c' in turn.

III. A Möbius transformation transforms circles into circles. (It is understood that straight lines come into the category of circles.)

We shall give two proofs of this important theorem.

In the first place let z be any point of the circle determined by the three points a, b, c. Then the cross-ratio

$$(za,bc) = \text{real number.}$$

Let z' be the transform of z. Then

$$(z'a',b'c') = (za,bc) = \text{real number,}$$

and therefore z' lies on a circle through a', b', c'.

The second proof will be given in the next section. We conclude this section with:

IV. A Möbius transformation conserves the angle of intersection of two circles.

Let b and c be the points common to two circles $\mathscr{C}_1$ and $\mathscr{C}_2$, a_1 a third point on $\mathscr{C}_1$ and a_2 a third point on $\mathscr{C}_2$. We have seen (see page 50) that the angle between the two circles is given by the argument of the cross-ratio (a_1a_2,bc). But this argument is conserved, since the cross-ratio itself is conserved.

A second proof of this theorem will also be given later.

7. A Möbius transformation dissected

The reader may ask, at this stage, whether there is any connection between Möbius transformations and inversion. We show that there is.

We may write

$$w = \frac{Az + B}{Cz + D} = \frac{A}{C} + \frac{BC - AD}{C(Cz + D)},$$

by simple division, or

$$w = \frac{A}{C} + \frac{BC - AD}{C^2} \cdot \frac{1}{z + D/C}.$$

We now carry out the transformation on z in several stages:

(i) Let $$z_1 = z + \frac{D}{C};$$

(ii) Let $\qquad z_2 = \dfrac{1}{z_1}\,;$

(iii) Let $\qquad z_3 = \dfrac{BC - AD}{C^2}\,.\,z_2\,;$

(iv) Then $\qquad w = \dfrac{A}{C} + z_3.$

Each stage can be interpreted geometrically, and this we proceed to do.

Stage (i) is equivalent to a *translation* of z in a definite direction and through a definite distance, given by the vector D/C.

Stage (ii) is equivalent to inversion in the unit circle, followed by reflection in the real axis. For with an obvious notation,

$$r_2(\cos \Theta_2 + i \sin \Theta_2) = \frac{1}{r_1(\cos \Theta_1 + i \sin \Theta_1)}$$

$$= \frac{1}{r_1} \left(\cos\left(-\Theta_1 \right) + i \sin \left(-\Theta_1 \right) \right),$$

and therefore $r_2 = 1/r_1$, so that $r_1 r_2 = 1$, and
$$\Theta_2 = -\Theta_1.$$

Stage (iii), *multiplication* by a fixed complex number is equivalent to a dilatation by a fixed amount from the origin, followed by a definite rotation about the origin. For if $(BC - AD)/C^2 = \rho(\cos \Theta + i \sin \Theta)$,

$$r_3 = \rho r_2,$$
and $\qquad\qquad \Theta_3 = \Theta_2 + \Theta.$

Stage (iv) is another translation.

We see that after each one of these sub-transformations circles are still circles (or straight lines), and that the angle of intersection of two circles is unchanged. Hence we have additional proofs of two of the theorems of the preceding section.

Another important deduction we can make from our dissection of a Möbius transformation is the following:

If a circle $\mathscr{C}$ is transformed into a circle $\mathscr{C}'$, the interior of $\mathscr{C}$ is mapped on the interior or the exterior of $\mathscr{C}'$.

We mean, of course, that every point inside $\mathscr{C}$ is mapped on points inside $\mathscr{C}'$, or every point inside $\mathscr{C}$ is mapped on points outside $\mathscr{C}'$.

The only transformation which can turn a circle inside-out is (ii), since this is an inversion followed by a reflection. The inverse of a circle with respect to a centre of inversion inside the circle turns the circle inside out. The other transformations which come into the dissection preserve insides and outsides.

If the circle $\mathscr{C}'$ is a straight line, the inside of $\mathscr{C}$ is mapped on one of the half-planes bounded by the line, the outside of $\mathscr{C}$ on the other.

8. The group property

The Möbius transformations of the z-plane form a group. Before we prove this, we give the definition of a *group of transformations*.

A set of transformations is said to form a group if it has the following properties:

(a) Every transformation T of the set possesses an inverse transformation which is also in the set. If this inverse transformation be denoted by T', the transformation effected by carrying out first T and then T' is the identical transformation, which leaves everything unchanged.

(b) If two transformations T_1, T_2 both belong to the set, the transformation obtained by carrying out first T_1 and then T_2 also belongs to the set.

That the Möbius transformations of the z plane form a group is a theorem easily proved by algebra. Alternatively we can proceed as follows:

The transformation inverse to
$$(za,bc) = (z'a',b'c'),$$
that is, the one which maps z' on z, is given by the equation
$$(z'a',b'c') = (za,bc).$$
Now let M_1 be a Möbius transformation which maps a on a', b on b', c on c'. It is determined by these three pairs of corresponding points. Let M_2 be a Möbius transformation which maps a' on a'', b' on b'' and c' on c''. It is also determined by these three pairs of corresponding points.

M_1 is given by the equation $(z'a',b'c') = (za,bc)$;

M_2 is given by the equation $(z''a'',b''c'') = (z'a',b'c')$.

The result of M_1 followed by M_2 is the transformation
$$(z''a'',b''c'') = (za,bc),$$
which is a Möbius transformation.

If a set of transformations, all of which belong to a group G of transformations, themselves form a group, we call this a *subgroup* of the group G. An important example of a group of transformations is afforded by the group of displacements in a Euclidean plane.

We imagine a fixed plane, and another like a large sheet of paper covering it. The paper is moved, with any possible figures on it, from one position over the plane to another. We consider only the initial and final positions of a motion or displacement. The path does not concern us. Suppose A, B, points of the

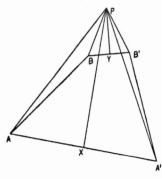

Fig. 54

plane in its first position, become A', B' in its final position. Join AA', BB', and let their perpendicular bisectors PX, PY meet in P. Evidently the triangles PAB, $PA'B'$ are congruent. Hence the displacement envisaged can be effected by a rotation round P.

If the right bisectors were parallel, then AA', BB' would be parallel, and the displacement would be a *translation*, every point of the plane moving through the same distance and in the same direction.

If we follow a given displacement by another, the result is a displacement, since only initial and final positions are considered. The inverse of a given displacement is merely that displacement which brings the moving plane back to its original position. Hence the displacements, or Euclidean motions of a plane, form a group.

The set of translations of the plane evidently form a subgroup of the group of displacements.

We may also regard the displacement of one plane over another as a *mapping* of the points of a Euclidean plane into itself which is one-to-one, transforms lines into lines, preserves angles, and also orientation.   It is this notion which is the basis of *applicability* in Euclidean geometry.   In the next section we investigate a group of transformations which, in our non-Euclidean geometry, corresponds to the group of Euclidean motions.

9. Special transformations

We are interested in a certain subgroup of the group of Möbius transformations, namely that subgroup which consists of all Möbius transformations which map the inside of a given circle on itself.   Let us be more explicit.

We choose the circle with centre at the origin and radius $= 1$. We call the circumference ω, and the circular area bounded by ω we call Ω.   The points of ω are not included in Ω.

There are Möbius transformations which transform Ω into itself. We shall call these M-transformations.   For example:

the identical transformation $w = z$;

the transformation $w = (\cos \varphi + i \sin \varphi)z$;

this gives a rotation of the plane, through an angle φ, about the origin.

If an M-transformation maps Ω on itself, the dissection of §7 shows that the boundary ω is also mapped on itself.   But the converse, as we know from the simple example $w = 1/z$, is not true.

It is easy to see that M-transformations form a group, which is a subgroup of the group of all Möbius transformations.   For, if a transformation maps Ω on itself, this is also true for the inverse transformation; and if two transformations M_1, M_2 both map Ω on itself, this is also true for the transformation obtained by first carrying out M_1 and then M_2.

We shall now restrict ourselves to these M-transformations and to Ω, the interior of the unit circle, centre at the origin.   We shall show that *Ω can be regarded as a non-Euclidean space, in which the M-transformations are the group of non-Euclidean motions.*

10. The fundamental theorem

We recall one of the characteristic properties of the group of displacements of a plane (over itself).   Let A be any point in the plane, α a direction through it, X any other point in the plane and

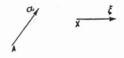

Fig. 55

$ξ$ a direction through it. Then there is a unique displacement of
the plane which maps A on X and α on $ξ$.

We show that the M-transformations have the same property in
$Ω$. The theorem to be proved is:

*Let X and A be any two points of $Ω$, $ξ$ and α directions through X
and A respectively. Then there is a unique M-transformation
which maps $(X,ξ)$ on $(A,α)$.*

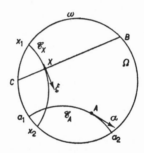

Fig. 56

We note in the first place that, through any point X, and
touching any given line through the point, there passes exactly
one circle orthogonal to ω. This is immediate by inversion, or
may be proved as follows: draw the perpendicular at X to the
given line, and let B, C be the points in which it cuts ω. Among
the circles of the pencil through B and C, two are evidently
orthogonal to the circle sought for: ω itself and the line BC.
Hence the circle required is of the coaxal system conjugate to
that defined by ω and the line BC. It therefore has B and C as
limiting points, and is uniquely determined, since it passes through
X.

If we care to regard it as an Apollonius circle, it is the locus of a
point which moves so that the ratio of its distances from two fixed
points B and C is always equal to XB/XC.

Call this circle $\mathscr{C}_X$, and let x_1 and x_2 be its intersections with ω. In the same way, there passes through A a unique circle $\mathscr{C}_A$ orthogonal to ω and tangent at A to the direction α. Let a_1, a_2 be its intersections with ω.

Now, since M-transformations map ω on itself, they necessarily map circles orthogonal to ω on to circles orthogonal to ω, being Möbius transformations.

Hence any M-transformation which maps (X,ξ) on (A,α) must necessarily map $\mathscr{C}_X$ on $\mathscr{C}_A$, since $\mathscr{C}_X$ will be mapped on a circle through A, with direction α, and orthogonal to ω, and there is only one such circle, namely $\mathscr{C}_A$.

Let us now suppose that the points of intersection of $\mathscr{C}_X$ and $\mathscr{C}_A$ with ω are so numbered that the direction ξ is that of x_1 towards x_2 in Ω, and the direction α is that of a_1 towards a_2 in Ω. The M-transformation we are looking for must map x_1 on a_1, x_2 on a_2 and X on A. These three pairs of corresponding points determine a unique Möbius transformation. Our proof will therefore be complete if we show it is an M-transformation, mapping Ω into itself. This is clear, since the unique Möbius transformation maps the circle through x_1 and x_2 orthogonal to $\mathscr{C}_X$, that is ω, on the circle through a_1 and a_2 orthogonal to $\mathscr{C}_A$, which is again ω. It therefore maps ω on ω. On the other hand it maps Ω on Ω, since it maps the interior point X of Ω on the point A of Ω. The unique Möbius transformation is therefore an M-transformation.

Corollary: Let a and b be two points of Ω, $\mathscr{C}_a$ a circle orthogonal to ω passing through a, $\mathscr{C}_b$ a circle orthogonal to ω passing through b. There exist precisely *two* M-transformations mapping a on b and $\mathscr{C}_a$ on $\mathscr{C}_b$. To a given orientation on $\mathscr{C}_a$ the one transforma-

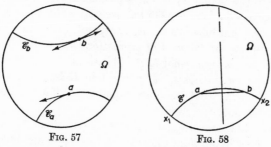

FIG. 57   FIG. 58

tion maps one of the possible directions of motion on $\mathscr{C}_b$, and the second transformation maps the other.

In particular, let a, b be two points of Ω. There exists a unique M-transformation which permutes a and b (that is, which maps a on b and b on a).

To prove this, we note in the first instance that there exists a unique circle orthogonal to ω and passing through a and b. Call this circle $\mathscr{C}$. An M-transformation interchanging a and b must transform $\mathscr{C}$ into itself.

By the above, there are only two M-transformations which map a on b and $\mathscr{C}$ on itself. The one which conserves orientation on $\mathscr{C}$ cannot map b on a, since then the sense of a towards b in Ω would be transformed into the sense of b towards a in Ω.

Let M' therefore be the M-transformation which maps a on b and alters the orientation of $\mathscr{C}$. We show that M' maps b on a.

Let x_1 and x_2 be the intersections of $\mathscr{C}$ and ω, and a' the transform of b by M'. We want to prove that $a' = a$. Since M' changes the orientation of $\mathscr{C}$, x_1 is mapped on x_2, x_2 on x_1. Since a is also mapped on b, and b on a', the cross-ratio property gives

$$(x_1 x_2, ab) = (x_2 x_1, ba').$$

But, for any cross-ratio,

$$(x_1 x_2, ab) = (x_2 x_1, ba).$$

Hence

$$(x_2 x_1, ba) = (x_2 x_1, ba'),$$

and therefore

$$a' = a.$$

11. The Poincaré model

We can now discuss Poincaré's model of a hyperbolic non-Euclidean geometry. The term "non-Euclidean" refers to the fact that in this geometry Euclid's parallel axiom does not obtain.

Our space is *the interior Ω of the circle* ω. We shall call this the h-plane. A point of Ω is a point, or h-point, in our new geometry. But *our h-lines are the arcs of circles (or straight lines) inside Ω, limited by ω and orthogonal to ω*. These are the arcs considered in our previous theorems. If a point in Ω lies on one of these arcs, we shall say that the corresponding h-point lies on the corresponding h-line.

We see that the ordinary incidence theorems of Euclidean geometry hold in our h-geometry. Through two distinct points of Ω there passes a unique circle orthogonal to ω. Hence, *through two distinct h-points there passes a unique h-line*.

Two circles orthogonal to ω intersect in at most one point which is inside Ω. Hence, two h-lines intersect in at most one h-point. We shall consider the pairs of h-lines which do not intersect later.

Since arcs of circles inside Ω orthogonal to ω constantly recur, we call them ω-arcs.

The angle between two ω-arcs will be, by definition, the h-angle between the corresponding h-lines.

The relation of equality between the angles of ω-arcs will imply the relation of equality, or h-equality, between h-angles.

We now come to the more difficult notion of h-*equality* between h-segments. An h-*segment* is naturally the portion of an ω-arc bounded by two points of Ω.

In elementary geometry two segments are equal if one can be applied to the other. This involves a displacement, or motion, of the plane. Such a motion transforms lines into lines, and conserves angles. In our h-geometry the equivalent to this group of motions is the group of M-transformations, which transform h-lines into h-lines, and conserve angles. We therefore call an M-transformation of Ω an h-*displacement of the h-plane*, and say that two h-segments AB and CD are h-equal, written

$$AB \overset{h}{=} CD,$$

if there is an h-motion which transforms AB into CD. Since M-transformations form a group, the relation of h-equality satisfies the usual axioms of equivalence:

(i) $AB \overset{h}{=} AB$:

(ii)   if $AB \overset{h}{=} CD$, then $CD \overset{h}{=} AB$:

(iii) if $AB \overset{h}{=} CD$, and $CD \overset{h}{=} EF$, then

$$AB \overset{h}{=} EF.$$

Again, since there is an M-transformation mapping the segment ab of an ω-arc on the segment ba, we also have

$$AB \overset{h}{=} BA.$$

The ordinary theorems on congruence in Euclidean geometry have an equivalent in our h-geometry. For example, let us prove the theorem:

In two h-triangles ABC and $A'B'C'$, the equalities $AB \overset{h}{=} A'B'$,

$AC \overset{h}{=} A'C'$ and $\angle CAB \overset{h}{=} \angle C'A'B'$ imply the equality $\angle ABC$

$\overset{h}{=} \angle A'B'C'$.

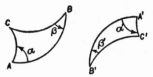

Fig. 59

To prove this we must show that if two triangles ABC, $A'B'C'$ (made up of ω-arcs) are such that $\alpha = \alpha'$ (Fig. 59), and there exists an M-transformation which maps AB on $A'B'$, and an M-transformation which maps AC on $A'C'$, then $\beta = \beta'$.

Suppose at first that the triangles ABC, $A'B'C'$ have the same orientation. The proof will consist in showing that there is an M-transformation which maps the ω-triangle ABC on the ω-triangle $A'B'C'$.

Since $AB \overset{h}{=} A'B'$, there exists a (unique) M-transformation which maps A on A' and B on B'. Since this transformation preserves angles in magnitude and orientation, it maps the ω-arc AC on the ω-arc $A'C'$ in the direction $A'C'$. This M-transformation is the only one which maps A on A' and the ω-arc AC in the direction AC on the ω-arc $A'C'$ in the direction $A'C'$. Hence, it is the one which maps A on A' and C on C', since we know such a transformation exists.

We have therefore found an M-transformation which maps the ω-triangle ABC on the ω-triangle $A'B'C'$. This transformation conserves angles, and therefore $\beta = \beta'$.

If the orientations of the ω-triangles ABC, $A'B'C'$ are different, we take *the geometrical image* of $A'B'C'$ in any diameter of Ω. Call the transformed triangle $A''B''C''$. Then the ω-triangles ABC, $A''B''C''$ have the same orientation, and the proof applies. Since the ω-triangle $A''B''C''$ arises from $A'B'C'$ by a reflection, the theorem follows.

12. The parallel axiom

So far we have shown that there is a complete equivalence between our *h*-geometry and that part of Euclidean geometry which is developed before the introduction of the parallel axiom. We now come to the parting of ways !

We could say that two ω-arcs which do not intersect shall be called *h-parallel h-lines.* Through a point *A* outside any ω-arc we can draw an infinity of ω-arcs which do not cut the given ω-arc. This would lead to there being an infinity of *h*-lines *h*-parallel

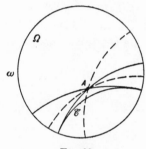

FIG. 60

to a given *h*-line through a given *h*-point outside the *h*-line.

We notice that there are two ω-arcs through *A* which *touch* the given ω-arc, the respective points of contact being on ω. These two ω-arcs separate the ω-arcs through *A* into two classes: those which cut the given ω-arc, and those which do not.

We prefer to reserve the term *h-parallel* for the two ω-arcs which touch the given ω-arc at a point of ω, and therefore now have, instead of Euclid's parallel axiom, the theorem:

Through any h-point A taken outside an h-line there pass precisely two h-parallels to the given h-line.

Hence, the term *non-Euclidean* geometry. The existence of a model for non-Euclidean geometry shows that the parallel axiom is *independent* of the other axioms, and *cannot be deduced from them.* Much fruitless effort has gone into attempts to *prove* the parallel axiom, and even today there exist organizations of people devoted to proving the axiom. Needless to say, these devotees are not shaken in their faith by mathematical arguments.

13. Non-Euclidean distance

We know how to compare two *h*-segments for equality. We

now want to obtain a *distance-function* $D[ab]$ for any two h-points a, b which satisfies the conditions:

$$D[ba] = D[ab] \geqslant 0;$$
$$D[ab] = 0 \quad \text{if and only if } b = a;$$
$$D[ab] + D[bc] \geqslant D[ac],$$

with equality if and only if b lies on the h-line ac between a and c.

We also want this distance function to be invariant under h-motions of the h-plane.

Let the h-line ab cut ω in the points α, β. Then

$$D[ab] = \left| \log (\alpha\beta, ab) \right|$$

is a suitable distance-function. It evidently satisfies the first two conditions, and also the condition that it be invariant under h-motions of the h-plane.

For $\quad D[ba] = \left| \log (\alpha\beta, ba) \right|$

$$= \left| \log \frac{1}{(\alpha\beta, ab)} \right| = \left| - \log (\alpha\beta, ab) \right|$$

$$= D[ab],$$

and, by definition, $D[ab] \geqslant 0$. It can only $= 0$ if $(\alpha\beta, ab) = 1$, when we must have $b = a$. Again, if any M-transformation of Ω maps the points α, β, a, b on α', β', a', b' respectively, we know that α', β' will be on ω, and that α', β', a', b' all lie on an h-line. Since also $(\alpha\beta, ab) = (\alpha'\beta', a'b')$,

$$D[ab] = D[a'b'].$$

If a, b, c are three points, in this order, on an h-line which cuts ω in α, β, we have, identically,

$$(\alpha\beta, ac) = (\alpha\beta, ab) . (\alpha\beta, bc).$$

As the points a, b, c are ordered as shown, the three logarithms $\log (\alpha\beta, ac)$, $\log (\alpha\beta, ab)$ and $\log (\alpha\beta, bc)$ all have the same sign. Therefore

$$D[ac] = D[ab] + D[bc].$$

To prove the full triangle inequality is more difficult, but we remark that since the parallel axiom is not used in Euclid's earlier theorems, we can apply the first twenty-eight propositions of Euclid's first book, without change, to our h-plane. We then obtain theorems concerning the congruence of triangles, as we have seen, the theorem that the greatest side of a triangle is opposite the greatest angle, and lastly, the triangle inequality:

$$D[ab] + D[bc] > D[ac],$$

if a, b, c are not, in this order, on an h-line.

It is possible to go much further into the study of non-Euclidean geometry, but we do not propose to do so in this book. The reader may find that his interest in the axioms of Euclidean geometry has been aroused, and he cannot do better than to study these afresh, and then see what are the essential differences between the non-Euclidean geometry we have been discussing and Euclidean geometry.

CHAPTER IV

This chapter discusses the solution of a classical problem: to show that, of all closed plane curves of a given perimeter, the circle encloses the greatest area. This property of a circle is usually referred to as its *isoperimetric* property.

1. Steiner's enlarging process

Steiner gave a simple geometric construction by which, given any closed convex plane curve K which is not a circle, it is possible to devise a new curve K^* which is also plane and closed, has the *same* perimeter, but contains a *greater* area than K.

It follows that K cannot be the solution of the isoperimetric problem.

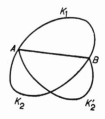

Fig. 61

The method is as follows. Choose on K two points A and B which bisect the perimeter; that is, the two arcs into which A and B divide K have equal length. If one arc be K_1, and if this, together with AB, bounds a curve of area F_1, while the other arc K_2 and AB bound a curve of area F_2, then we assume that $F_1 \geqslant F_2$. The area of the curve is $F = F_1 + F_2$.

We now wipe out the arc K_2, and substitute in its place the arc K_2' which is obtained from K_1 by reflection in the line AB. The closed curve bounded by K_1 and K_2' evidently has the same perimeter as K. Call this new curve K'. Its area is $F' = 2F_1$, and we have

$$F \leqslant F'.$$

Since the equality sign may hold in this relation, we have not yet completed the process. By hypothesis, K was not a circle. Hence, we may certainly choose A and B so that neither of the arcs K_1 and K_2 is a semicircle. Therefore K' is not a circle.

Hence we can find a point C distinct from A and B on K' so that $\angle ACB = \gamma$ is not a right angle. Let D be the image of C in AB. If we cut out from the area bounded by K' the quadrangle $ACBD$, four *lunes* remain, as indicated below.

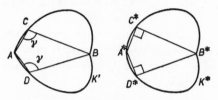

FIG. 62

We imagine that the sides of the quadrangle are bars, hinged at A, B, C and D, and that the lunes are fastened to these bars. We now move the bars about the hinges to a position where the angles at C and D are both right angles. Call the new quadrangle $A^*C^*B^*D^*$. Then the symmetric curve K^*, which circumscribes the new position of the quadrangle, is the one we are seeking.

In fact, the perimeter of K^* is equal to that of K', being made up of four arcs each congruent to the corresponding arc of K'. The difference in the areas enclosed by K' and K^* is equal to the difference in the areas of $ACBD$ and $A^*C^*B^*D^*$, since the lunes are unchanged in area. If F^* be the area bounded by K^*,

$$F^* - F' = CA.CB\,(1 - \sin \gamma) > 0.$$

Hence $F^* > F'$, and therefore

$$F^* > F,$$

so that our new curve K^* has the *same* perimeter, but encloses a *greater* area.

2. Existence of a solution

Does the process we have just described *prove* the isoperimetric property of the circle? We have shown that if K is a plane closed curve, not a circle, then the enlarging process always produces another closed plane curve K^* of the *same* perimeter but enclosing a *greater* area. Hence, the curve K cannot be a solution of the isoperimetric problem.

If, among all plane closed curves of a given perimeter, we can find one whose area $\geqslant$ the area of every other one, this curve must be a circle.

However, this does not solve the problem, since we must prove that such a curve exists. The Steiner construction indicates a method for proceeding *towards* a solution, but this is not enough, and we must digress for a moment.

If the perimeter of a given plane closed curve be L, we shall see later on that the area enclosed, F, satisfies the inequality

$$F < L^2.$$

Hence, the set of all plane closed curves with given perimeter L gives rise to a set of numbers F *bounded from above*. Such a set of numbers has a *least* upper bound $\varLambda$, say. Then $\varLambda$ has the following characteristic properties:

(i) $\varLambda \geqslant$ all numbers in the set;

(ii) for any $\epsilon > 0$, there exists at least one number in the set $> \varLambda - \epsilon$.

If the least upper bound $\varLambda$ is a *member* of the set of numbers F, the isoperimetric property of the circle is proved. But once again, this would have to be proved. The set of numbers $1/2$, $2/3, \ldots, n/(n+1), \ldots$ has 1 as a least upper bound, but 1 is not a member of the set, so that in a bounded infinite set of numbers there does not necessarily exist a greatest one.

We are therefore up against a serious difficulty, and, although it can be done, we shall not prove the isoperimetric property of the circle by a continued use of Steiner's enlarging process.

3. Method of solution

So far we have not discussed what we mean *precisely* by the perimeter of a closed curve, and the area enclosed by it. This remains to be done. What we shall do in this section is to enumerate the points in our proof of the isoperimetric property, giving the details later.

Let $\varLambda$ be the perimeter and $\varPhi$ the area of an equilateral polygon with an even number of vertices. Then we can show that

$$\varLambda^2 - 4\pi\varPhi > 0.$$

It follows from this that if K is any closed, rectifiable, continuous curve of perimeter L and area F, then

$$L^2 - 4\pi F \geqslant 0.$$

To show this, we inscribe in K an equilateral polygon V^* with an even number of vertices, and choose the sides ρ of V^* so small that the perimeter Λ^* and area Φ^* of V^* differ respectively from the perimeter L and area F of K by as small a quantity as we please. If we had

$$L^2 - 4\pi F < 0,$$

we could construct V^* so that

$$\Lambda^{*2} - 4\pi\Phi^* < 0,$$

and so arrive at a contradiction. Hence

$$L^2 - 4\pi F \geqslant 0.$$

If K is a circle, $L = 2\pi r$, $F = \pi r^2$, and

$$L^2 - 4\pi F = 0.$$

If K is any other continuous, closed and rectifiable curve, we can deduce from it, by the Steiner procedure, a curve K' of perimeter L' and area F' such that

$$L = L' \qquad \text{and} \qquad F < F'.$$

Since

$$L'^2 - 4\pi F' \geqslant 0,$$

it follows that we must have

$$L^2 - 4\pi F > 0.$$

Hence we have the following result:

Let K be a continuous, closed and rectifiable plane curve of perimeter L and area F. Then

$$L^2 - 4\pi F \geqslant 0,$$

and equality holds if and only if K is a circle, described positively.

This proves the isoperimetric property of the circle:

Amongst all plane curves of a given perimeter, the circle, described positively, has the greatest area.

The remainder of this chapter is devoted to the discussion of the concepts *perimeter* and *area* and to the proof of the statements made above.

4. Area of a polygon

Before we can discuss the concept *area enclosed by a curve,* we must be clear about the area of a polygon.

We take the usual system of rectangular Cartesian coordinates,

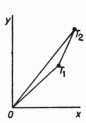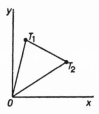

Fig. 63

and suppose that T_1, T_2 are two points with coordinates (x_1,y_1) and (x_2,y_2). Then the area of the triangle OT_1T_2 is given by the formula:

$$\text{area } OT_1T_2 = \tfrac{1}{2}(x_1y_2 - x_2y_1).$$

If we have $n + 1$ points: $T_i(x_i,y_i)$ $(i = 1, \ldots, n + 1)$, then, by the area of the *polygon* $OT_1T_2 \cdots T_{n+1}$ we understand the sum:

$$\text{area } OT_1T_2 + \text{area } OT_2T_3 + \cdots + \text{area } OT_nT_{n+1}$$

$$= \tfrac{1}{2} \sum_{k=1}^{n} (x_ky_{k+1} - x_{k+1}y_k).$$

We assume that T_{n+1} and T_1 coincide. The area is now

$$\tfrac{1}{2} \sum_{k=1}^{n} (x_ky_{k+1} - x_{k+1}y_k) \qquad (x_{n+1} = x_1; \; y_{n+1} = y_1).$$

A simple calculation shows that this is *independent of the position of the origin*, and we call it *the area of the closed polygon* $T_1T_2 \cdots T_n$.

By a *polygon*, we mean a finite number of points T_1, $T_2, \cdots, T_n$ which are not necessarily distinct, but are ordered. The area is also invariant under a rotation of the axes, and we can therefore say that *two similarly oriented congruent polygons have the same area*. If we reverse the sense in which the polygon is described, we change the sign of the area.

If two polygons have a sequence of vertices in common which are described in opposite senses for the two polygons, their areas can be added, and we obtain the area of the resulting polygon. For example:

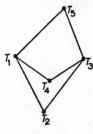

FIG. 64

$$\text{area } (T_1T_2T_3T_4) + \text{area } (T_1T_4T_3T_5) = \text{area } (T_1T_2T_3T_5).$$

The *perimeter* of a polygon is a simpler concept. We have

$$\text{perimeter } (T_1T_2 \cdots T_n) = \overset{n}{\underset{1}{\Sigma}} T_k T_{k+1},$$

where the length of each side is taken positively.

5. Regular polygons

We assume that we are given an equilateral polygon V with an even number of vertices n, and perimeter Λ. If n and Λ are fixed, when is the area of the polygon a maximum?

An equilateral polygon is said to be *regular* if its vertices lie on a circle and if, on traversing the circumference of the circle once, all the vertices are traversed just once.

We can show that the answer to the question posed above is: *the area of the polygon is a maximum when it is regular, and traversed positively.*

We prove this in two stages. By applying the Steiner enlarging process, we complete the first stage, and can show that *no polygon which is not regular can have a maximum area.*

The second stage is an existence proof: *Amongst all equilateral polygons with a given number n of vertices and a given perimeter Λ, there is one whose area $\geqslant$ area of any other.*

We shall leave the proof of the first stage to the reader, as it is quite straightforward. To prove the second stage, we first show that:

The areas of all the polygons we are considering are bounded from above.

We may assume that T_1 is at the origin. Then every length $OT_k \leqslant \Lambda/2$, for O and T_k are connected by two paths, made up of sides of the polygon, whose total length is Λ. One of these paths must therefore certainly be of length $\leqslant \Lambda/2$. Therefore the length of the straight segment $OT_k \leqslant \Lambda/2$, by the triangle inequality.

Since also $T_k T_{k+1} = \Lambda/n$, we have
$$|\text{ area of triangle } OT_k T_{k+1}| < \Lambda^2/4n.$$
On the other hand:
$$|\Phi| = |\text{ area } (T_1 T_2 \cdots T_n)| \leqslant n \max |\text{ area } (OT_k T_{k+1})|,$$
and therefore we have the desired result:
$$|\Phi| < \tfrac{1}{4}\Lambda^2.$$

The bounded set of the numbers denoting the areas Φ of all the polygons we are considering therefore has a least upper bound Φ_0, and we wish to show that Φ_0 is in the set; that is, that there exists at least one polygon with Φ_0 for the measure of its area.

To prove this, we invoke a classical theorem of function theory. If we consider the expression for the area
$$\Phi = \tfrac{1}{2} \sum_{k=1}^{n} (x_k y_{k+1} - x_{k+1} y_k),$$
it is evidently a continuous function of the $2n$ variables $x_1, y_1, \ldots, x_n, y_n$. If we keep T_1 at the origin, we can easily designate a closed area of the plane within which the polygon always lies. The square bounded by $x = -\Lambda, x = \Lambda, y = -\Lambda, y = \Lambda$ will do. Φ is a continuous function of $x_1, y_1, \ldots, x_n, y_n$ within and on the boundary of the product space given by:
$$x_1 = \pm \Lambda, y_1 = \pm \Lambda, \ldots, x_n = \pm \Lambda, y_n = \pm \Lambda.$$
But we also have the equations
$$(x_k - x_{k+1})^2 + (y_k - y_{k+1})^2 = \Lambda^2/n^2 \qquad (k = 1, \ldots, n).$$
In the product space these equations define a bounded, closed set of points. It is a classical theorem that Φ attains its maximum on this set.

Hence if Φ is the area of a *non-regular* equilateral polygon with an even number of vertices, and Φ_0 is the area of the *regular* polygon, described positively, with the same perimeter and number of sides, then
$$\Phi < \Phi_0.$$

If r is the radius of the circle circumscribing the regular polygon, then
$$\Lambda_0 = 2nr \sin \frac{\pi}{n},$$

$$\Phi_0 = n \left(\tfrac{1}{2} r^2 \sin \frac{2\pi}{n} \right) = nr^2 \sin \frac{\pi}{n} \cos \frac{\pi}{n},$$

so that

$$\Lambda_0^2 - \left(4n \tan \frac{\pi}{n} \right) \Phi_0 = 0.$$

For any equilateral polygon with n vertices and perimeter $\Lambda = \Lambda_0$, we have, by the above proof,

$$\Lambda^2 - \left(4n \tan \frac{\pi}{n} \right) \Phi > 0,$$

so that, in all cases,

$$\Lambda^2 - \left(4n \tan \frac{\pi}{n} \right) \Phi \geqslant 0.$$

We replace this by a weaker inequality which does not involve the number of sides.

Write $\dfrac{\pi}{n} = p$, then

$$4n \tan \frac{\pi}{n} = 4\pi \tan p / p.$$

But since $0 < p < \dfrac{\pi}{2}$, we know that $\tan p > p$, so that

$$4\pi \tan p / p > 4\pi,$$

and our inequality becomes

$$\Lambda^2 - 4\pi\Phi > 0.$$

This is the fundamental inequality for an equilateral polygon with an even number of sides used in §3.

6. Rectifiable curves

We now define the *perimeter* of a closed curve.

If $a \leqslant t \leqslant b$, the continuous functions $x(t)$, $y(t)$ give a parametric representation

$$x = x(t), \, y = y(t)$$

of a *continuous curve K* with initial point A ($t = a$) and end point B ($t = b$). We suppose that there is no subinterval $\alpha \leqslant t \leqslant \beta$ in which both functions x, y are constant.

We now take on K the points $T_1, T_2, \ldots, T_{n-1}$ which correspond to the parametric values

$$a < t_1 < t_2 < \cdots < t_{n-1} < b.$$

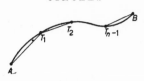

FIG. 65

We join the points $A, T_1, T_2, \ldots, T_{n-1}, B$ by straight lines in order, and obtain the sum
$$\Lambda = AT_1 + T_1T_2 + \cdots + T_{n-1}B,$$
giving the length of the inscribed path $AT_1T_2 \ldots T_{n-1}B$.

If this sum is always bounded, we say that K is *rectifiable,* and the least upper bound of the set of positive numbers Λ is called the *arc length L* of K.

From the property of the least upper bound, we always have
$$\Lambda \leqslant L,$$
and for any $\epsilon > 0$ there always exist values Λ satisfying
$$\Lambda > L - \epsilon.$$

An immediate deduction from the definition of arc length is the following: if two points A and B are joined by a continuous rectifiable curve K which is distinct from the line AB, then the arc length of K exceeds the length AB.

For if K contains a point M which is not on the line AB, then the arc length L of K satisfies the inequality
$$L \geqslant AM + MB > AB.$$

If M is a point on K corresponding to a value m, where $a < m < b$, it can be proved, in an evident notation, that
$$L_a^b = L_a^m + L_m^b.$$
We leave this proof to the reader.

We now consider a *closed* continuous curve K:
$$x = x(t), \qquad y = y(t), \qquad (a \leqslant t \leqslant b),$$
$$x(a) = x(b), \qquad y(a) = y(b).$$
We can remove the restriction that $a \leqslant t \leqslant b$ if we decide that $x(t), y(t)$ are to have the *period* $b - a$; that is,
$$x(t + b - a) = x(t),$$
$$y(t + b - a) = y(t),$$
for all values of t. By a linear substitution,
$$\varphi = pt + q,$$
we can transform the interval $a \leqslant t \leqslant b$ into the interval $0 \leqslant \varphi \leqslant 2\pi$. Replacing t by φ, we have two continuous functions:

$$x = x(\varphi), \qquad y = y(\varphi),$$

of period 2π. If, besides this, we put

$$\xi = \cos\varphi, \qquad \eta = \sin\varphi,$$

(ξ, η) describes the unit circle. Every point of this circle corresponds, to a multiple of 2π, to a unique value of φ, and this value of φ corresponds to a unique point on K. Hence we may say:

By a continuous closed curve we mean the continuous mapping of the points of a circle. Such a mapping is not necessarily one-to-one.

We now define the *perimeter* of a closed curve. We merely use the previous definition of arc length when $A = B$:

$$L = L_a^b.$$

L is independent of the choice of the point $A = B$; that is,

$$L_a^b = L_{a+c}^{b+c}.$$

By the rule for the addition of arcs, to prove this we must show that

$$L_a^{a+c} + L_{a+c}^b = L_{a+c}^b + L_b^{b+c},$$

which reduces to

$$L_a^{a+c} = L_b^{b+c}.$$

Since $a = b$, this is evident.

We can sum up for a closed curve K and say:

Inscribe in K a polygon whose vertices follow each other in the correct cyclic order. If the least upper bound L of the perimeters Λ of all these inscribed polygons is finite, we say that K is rectifiable, and L is its perimeter.

7. Approximation by polygons

We return once again to a rectifiable arc K:

$$x = x(t), \qquad y = y(t), \qquad (a \leqslant t \leqslant b),$$

whose initial point A and end point B do not necessarily coincide. We show that the arc length L of K can be approached by the lengths Λ of inscribed straight line paths in a *uniform* manner. Let V be such a path with vertices corresponding to the parametric values

$$a = t_0 < t_1 < t_2 < \cdots < t_n = b.$$

We have the following theorem:

Given any $\epsilon > 0$ we can always determine $\delta > 0$ so that the length Λ of an inscribed path V differs from the length L of K by less than ϵ, that is

$$L - \Lambda < \epsilon,$$

as soon as all the parametric differences

$$t_k - t_{k-1} < \delta \qquad (k = 1, \ldots, n).$$

From the definition of L we can find a path V' inscribed in K whose length Λ' satisfies the inequality

$$L - \Lambda' < \epsilon/2.$$

Let $A = T_0, T_1', \ldots, T_m' = B$ be the vertices of V'. If we choose δ so that

$$\delta < t_k' - t_{k-1}' \qquad (k = 1, \ldots, m),$$

and inscribe in K a path V for which

$$t_k - t_{k-1} < \delta \qquad (k = 1, \ldots, n),$$

then, between two consecutive vertices T_{k-1}', T_k' of V' there is at least one vertex T_r of V. The path from A to B which contains the vertices of V and of V' has a length Λ'' such that

$$\Lambda'' \geqslant \Lambda', \text{ and therefore } L - \Lambda'' < \epsilon/2.$$

Suppose now that T_k' lies between T_r and T_{r+1}. Since the continuity of $x(t)$, $y(t)$ implies uniform continuity, for a suitably small value of δ both the distances $T_r T_k'$ and $T_k' T_{r+1}$ will be smaller than an arbitrarily given $\eta > 0$. Then

$$\Lambda'' - \Lambda = \Sigma(T_r T_k' + T_k' T_{r+1} - T_r T_{r+1})$$
$$< \Sigma(T_r T_k' + T_k' T_{r+1}) < 2\eta m.$$

Hence

$$L - \Lambda < \frac{\epsilon}{2} + 2\eta m.$$

We can choose δ so that

$$\eta < \frac{\epsilon}{4m},$$

and we then have

$$L - \Lambda < \epsilon,$$

which is our goal.

We can make the successive segments of our paths equal, if we wish. Describe a circle with A, the initial point of K, as centre, and radius ρ so small that K does not lie entirely within this circle. The first point of K, that is the point which has the smallest t-value and lies on the circumference of the circle, we call T_1. With T_1 as centre and radius ρ we describe a circle, and call the first point following T_1 which lies on its circumference T_2. We

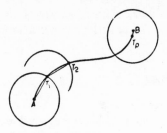

Fig. 66

continue in this way, and obtain a path consisting of equal segments of length ρ inscribed in K.

As the length of this path cannot exceed L, after a finite number of steps we reach a point T_p such that the arc of K between T_p and B lies within the circle radius ρ and centre T_p. If we now join up the points $A, T_1, T_2, \ldots, T_p, B$ in order, we obtain a path V^* inscribed in K whose first p sides are of length ρ, whilst the last side $\leqslant \rho$. We now prove:

ρ can be taken so small that all the parametric differences $t_{k+1} - t_k$ $(k = 0, 1, \ldots, p)$ corresponding to the vertices $A = T_0, T_1, \ldots,$ $T_p, T_{p+1} = B$ are smaller than an arbitrary positive number δ:

$$t_{k+1} - t_k < \delta \qquad (k = 0, 1, \ldots, p).$$

To see this, we first construct a path V' with vertices $A = T_0', T_1', \ldots, T_{m+1}' = B$ for which $t_{k+1}' - t_k' < \delta/2$. Since, by hypothesis, no subinterval $\alpha \leqslant t \leqslant \beta$ corresponds to a single point of K, we can choose V' so that no two consecutive points T_k' and T_{k+1}' coincide. If we now take 2ρ smaller than the smallest segment of V', there must lie between two successive vertices T' at least one vertex T, and therefore all differences $t_{k+1} - t_k < \delta$, as desired.

Since $T_p B \leqslant \rho$, the path V^* is not, in general, equilateral. If we replace this last segment by a path (not inscribed) of two or three sides each of length ρ, we obtain, instead of V^*, an equilateral path $'V^*$ with an *even* number of sides. This path is not inscribed, in the previous sense.

But if we take $A = B$, so that K is closed, the difference between the areas of the polygons V^* and $'V^*$ is equal to the area of the triangle (or quadrangle) constructed on $T_p B$ as base, and is therefore $< 4\rho^2$, by a previous result (see page 70), whilst the perimeters differ by less than 3ρ. Hence we can say:

Given $\epsilon > 0$ we can determine ρ, the side of an equilateral polygon V^* with an even number of vertices approximating to K so that the relation

$$L - \Lambda^* < \epsilon$$

holds between the length of K and the perimeter of V^*.

8. Area enclosed by a curve

Once again we assume K to be closed, continuous and rectifiable:

$$x = x(t), \qquad y = y(t) \qquad (a \leqslant t \leqslant b),$$
$$x(a) = x(b), \qquad y(a) = y(b).$$

We inscribe in K a polygon $A = T_0, T_1, \ldots, T_n = B = A$, corresponding to parametric values

$$a = t_0 < t_1 < \cdots < t_n = b.$$

The area Φ of this polygon is given by the formula:

$$2\Phi = \sum_{k=1}^{n} \left[x(t_{k-1})y(t_k) - y(t_{k-1})x(t_k) \right].$$

We wish to show that, given any $\epsilon > 0$ we can find $\delta > 0$ such that if

$$t_k - t_{k-1} < \delta \qquad (k = 1, \ldots, n),$$

then $$F - \Phi < \epsilon,$$

where F is a number which we shall call *the area enclosed by K*.

We cannot do this without a further study of the functions $x(t), y(t)$. We know that K is rectifiable. Hence

$$\sum_{k=1}^{n} \left[\{x(t_k) - x(t_{k-1})\}^2 + \{y(t_k) - y(t_{k-1})\}^2 \right]^{\frac{1}{2}}$$

is bounded, for all possible polygons inscribed in K. Since

$$[\{x(t_k) - x(t_{k-1})\}^2 + \{y(t_k) - y(t_{k-1})\}^2]^{\frac{1}{2}}$$
$$\geqslant \left| x(t_k) - x(t_{k-1}) \right|,$$
$$\geqslant \left| y(t_k) - y(t_{k-1}) \right|,$$

we see that the function $x(t)$ has the property that the sum

$$\sum_{k=1}^{n} \left| x(t_k) - x(t_{k-1}) \right|$$

is always bounded. Similarly for $y(t)$. Such functions are called *functions of bounded variation*. Hence, if K is rectifiable, the continuous functions $x(t), y(t)$ must both be of bounded variation. Since

$$[\{x(t_k) - x(t_{k-1})\}^2 + \{y(t_k) - y(t_{k-1})\}^2]^{\frac{1}{2}}$$
$$\leqslant \left| x(t_k) - x(t_{k-1}) \right| + \left| y(t_k) - y(t_{k-1}) \right|,$$

the condition is also sufficient.

To facilitate the handling of the expression for Φ, we prove the following theorem:

Every continuous function $f(t)$ of bounded variation can be written as the difference

$$f(t) = \varphi(t) - \psi(t)$$

of two continuous and monotone increasing functions.

Let $\alpha \leqslant t \leqslant \beta$ be any subinterval of $a \leqslant t \leqslant b$. We divide it up as shown:

$$\alpha = t_0 < t_1 < \cdots < t_n = \beta,$$

and form the sum

$$\sum_{k=1}^{n} \left| f(t_k) - f(t_{k-1}) \right|.$$

By hypothesis, this sum is bounded. Let the least upper bound for all subdivisions of $\alpha \leqslant t \leqslant \beta$ be denoted by

$$S_\alpha^\beta f.$$

Actually, this is the arc-length of the continuous curve

$$x = f(t), \qquad y = 0$$

which lies on the x-axis, between $t = \alpha$ and $t = \beta$.

Now

$$\varphi(t) = S_\alpha^t$$

is a monotone increasing, continuous function. The monotony arises from the additive property of arc lengths:

$$S_\alpha^t + S_t^{t+h} = S_\alpha^{t+h}.$$

Again, from the first theorem of §7, we have, for $h < \delta$,

$$S_t^{t+h} - \left| f(t+h) - f(t) \right| < \epsilon,$$

since the first term denotes an arc-length, and the second the length of a one-sided inscribed path. Therefore

$$\varphi(t+h) - \varphi(t) = S_t^{t+h} < \epsilon + \left| f(t+h) - f(t) \right|,$$

and because of the continuity of f, the right-hand side can be made as small as we please by diminishing h. Hence $\varphi(t)$ is continuous.

The continuous function

$$\psi(t) = \varphi(t) - f(t)$$

is also monotone increasing. For by the definition of φ, for $\alpha < \beta$,

$$\varphi(\beta) - \varphi(\alpha) = S_\alpha^\beta \geqslant \left| f(\beta) - f(\alpha) \right|.$$

We therefore have the desired representation:

$$f(t) = \varphi(t) - \psi(t).$$

Returning to the formula for Φ, we represent $y(t)$, which is of bounded variation, in this way:

$$y(t) = \varphi(t) - \psi(t).$$

Then
$$2\Phi = \sum_{k=1}^{n}\left[x(t_{k-1})y(t_k) - y(t_{k-1})x(t_k)\right]$$
$$= \sum_{k=1}^{n} x(t_{k-1})\Big\{y(t_k) - y(t_{k-1})\Big\} - \sum_{k=1}^{n} y(t_{k-1})\Big\{x(t_k) - x(t_{k-1})\Big\}.$$

Now
$$\sum_{k=1}^{n} x(t_{k-1})\Big\{y(t_k) - y(t_{k-1})\Big\} = \sum_{k=1}^{n} x(t_{k-1})\Big\{\varphi(t_k) - \varphi(t_{k-1})\Big\}$$
$$- \sum_{k=1}^{n} x(t_{k-1})\Big\{\psi(t_k) - \psi(t_{k-1})\Big\}.$$

Both sums on the right-hand side are in a form to which the usual existence proof for the Riemann definite integral applies. This shows that each sum:

$$\sum_{k=1}^{n} x(t_{k-1})\Big\{\varphi(t_k) - \varphi(t_{k-1})\Big\},$$

$$\sum_{k=1}^{n} x(t_{k-1})\Big\{\psi(t_k) - \psi(t_{k-1})\Big\}$$

tends to a definite limit as we refine the subdivisions of the interval $a \leqslant t \leqslant b$; that is, as we diminish δ. A similar result holds for

$$\sum_{k=1}^{n} y(t_{k-1})\Big\{x(t_k) - x(t_{k-1})\Big\},$$

as we see by representing $x(t)$ as the difference of two monotone increasing continuous functions.

Hence we finally have the theorem:

If $\epsilon > 0$ we can choose the length ρ of the side of an equilateral polygon V^* approximating to K so that between the area F enclosed by K and the area Φ^* enclosed by V^* the inequality
$$|F - \Phi^*| < \epsilon$$
holds. V^* can be chosen to have an even number of vertices.

This completes the proof of the theorems needed in §3.

Naturally the circle is a solution of many other problems in mathematics. We mention only one. If we have an enclosed plane membrane of given area, and set it vibrating, what form must the boundary take for the fundamental tone to be as deep as possible ? The answer is a circle.

It is evident that if we pursued circles conscientiously, very few branches of pure or applied mathematics would be left unexplored.

EXERCISES

1.1. With the notation of p. 1 for a triangle ABC, and the use of the power concept of p. 15, prove that $CA'.CP = CB'.CQ$, so that a circle $\mathscr{C}_3$ passes through the points A', P, B', Q. Similarly, prove that a circle $\mathscr{C}_2$ passes through the points C', R, A', P. If $\mathscr{C}_2 \neq \mathscr{C}_3$, what is their common chord? Using the theorem on radical axes of p. 21, obtain another proof of the nine-point circle theorem.

1.2. Using compasses only, determine when the construction for the inverse of a point (p. 5) breaks down. Using compasses only, show that we can find a point Y on OX such that $OY = nOX$ (integral n) (see p. 24). If n is large enough, we can find the inverse Y' of Y. Show how this enables us to find the inverse of X, when the construction for X' does not apply.

1.3. P, Q are points which are inverse in a given circle $\mathscr{C}$, and the figure is inverted with respect to any given circle Σ, so that we obtain a circle $\mathscr{C}'$ and points P' and Q'. Prove that P' and Q' are inverse points with respect to the circle $\mathscr{C}'$. What happens if $\mathscr{C}'$ is a line?

1.4. $\mathscr{C}$ and $\mathscr{D}$ are distinct circles. Find the locus of a point P which is such that two circles, each touching $\mathscr{C}$ and $\mathscr{D}$, also touch at P. (There are three cases to examine: $\mathscr{C}$ intersects $\mathscr{D}$, $\mathscr{C}$ touches $\mathscr{D}$, and $\mathscr{C}$ does not intersect $\mathscr{D}$.)

1.5. Prove that two circles $\mathscr{C}$ and $\mathscr{D}$ which intersect at B and C are orthogonal if and only if there exist two circles touching $\mathscr{D}$ at B and at C respectively which also touch each other at A, where A is any point on $\mathscr{C}$ distinct from B and C.

1.6. Let $\mathscr{C}$ and $\mathscr{D}$ be circles which intersect in the points A and B. By inverting with A as centre, show that the operations of inversion in $\mathscr{C}$ and $\mathscr{D}$ are commutative if and only if $\mathscr{C}$ and $\mathscr{D}$ are orthogonal. (If inversion in $\mathscr{C}$ is represented by $T_{\mathscr{C}}$, and inversion in $\mathscr{D}$ is represented by $T_{\mathscr{D}}$, we wish to prove that if P is any point then $(P)T_{\mathscr{C}}T_{\mathscr{D}} = (P)T_{\mathscr{D}}T_{\mathscr{C}}$ if and only if $\mathscr{C}$ and $\mathscr{D}$ are orthogonal. The theorem of Exercise 1.3 is to be used.)

1.7. In Exercise 1.6, above, can inversion be commutative if $\mathscr{C}$ and $\mathscr{D}$ touch, or if $\mathscr{C}$ and $\mathscr{D}$ do not intersect?

1.8. In triangle ABC, $\angle BAC = \alpha > 120°$. P is any point inside the triangle distinct from A. Let $\beta = 180° - \alpha$, so that $\beta < 60°$. Rotate triangle ABP about the vertex A through $\beta°$ to triangle $AB'P'$, A remaining fixed, B mapping on B' and P on P'. Then $\angle BAB' = \beta$, and since $\beta + \angle BAC = 180°$, the points B', A, C are collinear. Prove that in the isosceles triangle PAP' we have $AP > PP'$, and that

$$AP + PB + PC > PP' + P'B' + PC = B'P' + P'P + PC$$
$$\geqslant B'C = B'A + AC = AB + AC,$$

so that A is the Fermat point in this case (p. 12). (This proof is due to Dan Sokolowsky.)

1.9. Show that the power of a point P with respect to a point-circle (one of zero radius) centre A is equal to $(PA)^2$. Hence determine the locus of a point P which moves so that the ratio $PA:PB$ of its distances from two given points A and B is constant.

1.10. A is a point of intersection of circles $\mathscr{C}$ and $\mathscr{D}$, and a line through A intersects $\mathscr{C}$ again in P, and $\mathscr{D}$ again in Q. If V is the midpoint of PQ, prove that as the line through A varies, the point V moves on a circle. (Compare the powers of V with respect to $\mathscr{C}$ and to $\mathscr{D}$.) Examine the case when V divides PQ in a given ratio.

1.11. QR is a chord of a circle $\mathscr{C}$ which subtends a right angle at a given point L (the angle QLR is a right angle). If P is the midpoint of QR, prove that the power of P with respect to $\mathscr{C}$ is $-(PL)^2$. Deduce that the locus of P as the chord QR varies is a circle in the pencil determined by $\mathscr{C}$ and the point-circle L.

1.12. Prove that the coaxal system of circles $\lambda C + \mu C' = 0$, where $C \equiv 3(x^2 + y^2) - 2x - 4y + 3 = 0$, $C' \equiv 3(x^2 + y^2) - 4x - 2y + 3 = 0$ contains two point-circles, and that the equation of the coaxal system may be written in the form:

$$k_1[(x - 1)^2 + y^2] + k_2[x^2 + (y - 1)^2] = 0.$$

1.13. If $\mathscr{C}$ and $\mathscr{C}'$ are non-intersecting circles, and $\mathscr{C}_1$, $\mathscr{C}_2$, $\mathscr{C}_3, \ldots, \mathscr{C}_n$ touch $\mathscr{C}$ and $\mathscr{C}'$ and also each other as in Fig. 25 (p. 20), prove that the points of contact of the circles $\mathscr{C}_i$ with each other lie on a circle $\mathscr{D}$, and that the inverse of $\mathscr{C}$ in $\mathscr{D}$ is the circle $\mathscr{C}'$.

1.14. $\mathscr{C}_1$ and $\mathscr{C}_2$ are point-circles, and $\mathscr{C}_3$ is a proper circle. Prove that there are two circles which pass through the point-circles and touch $\mathscr{C}_3$.

1.15. $\mathscr{C}_1$ is a point-circle, $\mathscr{C}_2$ and $\mathscr{C}_3$ are proper circles which touch each other. Prove that there are three solutions to the Apollonian problem.

1.16. $\mathscr{C}_1$, $\mathscr{C}_2$ and $\mathscr{C}_3$ are proper circles, and $\mathscr{C}_1$ touches both $\mathscr{C}_2$ and $\mathscr{C}_3$, but these two do not touch each other. Prove that there are four solutions to the Apollonian problem, and five if we include $\mathscr{C}_1$.

1.17. $\mathscr{C}_1$ and $\mathscr{C}_2$ touch, and $\mathscr{C}_3$ is a proper circle which does not touch either. Prove that there are six distinct circles which touch the three given circles.

1.18. Using compasses only, justify the following construction for determining the centre of a given circle $\mathscr{C}$. With centre at any point A of $\mathscr{C}$, draw a circle $\mathscr{D}$ to cut $\mathscr{C}$ at points B and C. Find the geometric image O' of A in the line BC. Then the inverse of O' in the circle $\mathscr{D}$ is the centre of the circle $\mathscr{C}$.

1.19. A triangular cut-out ABC is given and we move it so that the side AB always passes through a given point L, and the side AC always passes through another given point M. Prove that the side BC of the cut-out touches a fixed circle.

1.20. *The Bobillier Envelope Theorem.* A triangular cut-out ABC is given, and we move it so that AB always touches a given circle $\mathscr{C}$, and AC always touches another given circle $\mathscr{D}$. Prove that the side BC of the cut-out touches a fixed circle $\mathscr{E}$.

Chapter II

2.1. Show that a line l which lies in the Oxy-plane in E_3 represents a coaxal system of circles which all pass through the point $(0,0)$. Where is the other common point of intersection? What kind of coaxal system is represented by l if l passes through $(0,0)$ and lies in the Oxy-plane?

2.2. Show that the coaxal system determined by the two circles:

$$C \equiv x^2 + y^2 - 2px - 2qy - k^2 = 0,$$
$$C' \equiv x^2 + y^2 - 2p'x - 2q'y - k^2 = 0$$

is always of the intersecting type, with distinct points of intersection if $k \neq 0$.

2.3. Show that two coaxal systems of circles in the Oxy-plane whose maps in E_3 are parallel lines are coaxal systems with the same radical axis.

2.4. Verify that for all a, b, and $c \neq 0$, every circle of the system:

$$a(x - x') + b(y - y') + c(x^2 + y^2 - z') = 0$$

intersects the fixed circle of the system:

$$-2x'(x - x') - 2y'(y - y') + x^2 + y^2 - z' = 0$$

at the ends of a diameter of this fixed circle.

2.5. A plane in E_3 which contains the point (p,q,r) has the equation:

$$a(x - p) + b(y - q) + c(z - r) = 0.$$

Deduce that the circle $a(x - p) + b(y - q) + c(x^2 + y^2 - r) = 0$ is always orthogonal to the circle $x^2 + y^2 - 2px - 2qy + r = 0$.

2.6. Prove that the circle which is orthogonal to each of three given circles $\mathscr{C}_1$, $\mathscr{C}_2$, $\mathscr{C}_3$ is uniquely defined, unless the three given circles belong to a coaxal system. If this is the case, are there any circles which touch the three given circles?

2.7. If three given circles are not in a coaxal system, and $\mathscr{C}_0$ is the uniquely defined common orthogonal circle (Exercise 2.6, above), show that the eight Apollonian contact circles (p. 21) are inverse in pairs in $\mathscr{C}_0$.

Chapter III

3.1. Prove that any Möbius transformation with the two distinct fixed points p and q may be written in the form:

$$(w - p)/(w - q) = k(z - p)/(z - q),$$

where k is a complex number. (A fixed point under a Möbius transformation is one which is mapped onto itself.)

3.2. Show that the points of any given circle in the z-plane may be represented thus: $z = (At + B)/(Ct + D)$, where A, B, C, D are complex numbers with $AD - BC \neq 0$, and t is real.

3.3. If ABC is an h-triangle in the Poincaré model (p. 58), where A is at the centre of ω, so that AB and AC are diameters of ω, show that BC is convex to A, and that the sum of the angles of the h-triangle ABC is less than the sum of the angles of the Euclidean triangle ABC, which is 180°.

3.4. If ABC is any h-triangle in Ω, prove that the angle sum is less than 180°. (Find an inversion which maps ω into itself and A into the centre.)

3.5. Show that through any given h-point A there exists a unique h-line l' perpendicular to a given h-line l, and that if m, n are the h-lines through A which are h-parallel to l, it being assumed that A does not lie on l, then m and n make equal angles with l'.

3.6. Prove that two distinct h-lines which are each perpendicular to the same h-line have no intersection.

3.7. Prove that two distinct h-lines which have no intersection have a unique h-line perpendicular to both of them.

SOLUTIONS

CHAPTER I

1.1. Circle on AB as diameter through P, Q. Therefore $CP.CB = CQ.CA$, but $CB = 2CA'$, $CA = 2CB'$. Common chord of $\mathscr{C}_2$, $\mathscr{C}_3$ is BC. Common chords of three circles cannot form sides of triangle. Hence one circle through A', P, R, C', Q, B'. Apply this theorem to triangle BHC, which shows above circle also goes through midpoints of HA, HB and therefore of HC.

1.2. If $OX < \frac{1}{2}k$, choose n so that $OY = nOX > \frac{1}{2}k$, and find Y', where $OY.OY' = k^2 = (nOX)(OY')$. Find X', where $X' = nOY'$, then $OX.OX' = k^2$.

1.3. All circles $\mathscr{D}$ through P, Q are orthogonal to $\mathscr{C}$. Therefore all circles $\mathscr{D}'$ through P', Q' are orthogonal to $\mathscr{C}'$. Hence P', Q' inverse in $\mathscr{C}'$. If $\mathscr{C}'$ is a line, P', Q' are mirror images in line.

1.4. If $\mathscr{C}$, $\mathscr{D}$ intersect at A, B, take A as centre of inversion. Locus of P' consists of two angle-bisectors of lines $\mathscr{C}'$, $\mathscr{D}'$ intersecting at B', so locus of P consists of two circles through A, B. If $\mathscr{C}$, $\mathscr{D}$ touch at A, $\mathscr{C}'$, $\mathscr{D}'$ are parallel lines, P' moves on halfway parallel line, and P moves on circle which touches both $\mathscr{C}$ and $\mathscr{D}$ at A. If $\mathscr{C}$, $\mathscr{D}$ do not intersect, invert into concentric circles, and P' moves on a third concentric circle, and therefore P moves on a circle in coaxal system determined by $\mathscr{C}$ and $\mathscr{D}$.

1.5. Invert, centre B, then $\mathscr{C}'$, $\mathscr{D}'$ are lines intersecting at C'. Circle touching $\mathscr{D}$ at B inverts into line parallel to $\mathscr{D}'$, intersecting $\mathscr{C}'$ at A'. Circle touching $\mathscr{D}$ at C inverts into circle touching $\mathscr{D}'$ at C' and through A'. The line through A' parallel to $\mathscr{D}'$ touches this latter circle at A' if and only if $\mathscr{C}'$ is orthogonal to $\mathscr{D}'$.

1.6. Take A as centre of inversion, then $\mathscr{C}'$ and $\mathscr{D}'$ are lines intersecting at B', and inversions in $\mathscr{C}$ and $\mathscr{D}$ become line reflections in $\mathscr{C}'$ and $\mathscr{D}'$. These are commutative if and only if $\mathscr{C}'$ is perpendicular to $\mathscr{D}'$, since a composition of reflections in two intersecting lines is equivalent to a rotation about their intersection through twice the angle between the lines, measured from the line of the first reflection to the second line.

1.7. No! Composition of reflections in parallel lines (after inversion) is equivalent to a translation, directed from the first line to the second, and composition of inversions in concentric circles (after inversion of the two non-intersecting circles) is equivalent to an enlargement with scale-factor depending on the order of inversion.

1.8. In a triangle the greatest side is opposite the greatest angle, and $\angle PP'B > 60° > \beta$. The shortest distance from B' to C is $= B'C$.

1.9. The equation to a point-circle (α, β) is $(x - \alpha)^2 + (y - \beta)^2 = 0$, and the power of $P(x,y)$ is $(x - \alpha)^2 + (y - \beta)^2 = (PA)^2$. Locus is circle of coaxal system determined by point-circles A and B (p. 15).

1.10. Power of V with respect to $\mathscr{C} = VA.VP$, and with respect to $\mathscr{D} = VA.VQ$. If $VP = -VQ$, then ratio of powers $= -1$, so by theorem on p. 15 the point V moves on circle through intersections of $\mathscr{C}$ and $\mathscr{D}$. Similar result if V divides PQ in a given ratio.

1.11. $PQ = PR = PL$, and so power of P with respect to $\mathscr{C} = -(PQ)^2 = -(PL)^2$. Power of P with respect to point-circle $L = (PL)^2$, so ratio of powers $= -1$, and theorem on p. 15 gives result.

1.12. Find the radius of $\lambda C + \mu C' = 0$, and equate to zero, obtaining a quadratic equation giving point-circles $(x - 1)^2 + y^2 = 0$, $x^2 + (y - 1)^2 = 0$, and use result at bottom of p. 15.

1.13. Invert $\mathscr{C}$, $\mathscr{C}'$ into concentric circles $\mathscr{E}$, $\mathscr{E}'$, centre V, and suppose that $\mathscr{C}_i$, $\mathscr{C}_{i+1}$ invert into $\mathscr{E}_i$, $\mathscr{E}_{i+1}$ which touch each other at T, $\mathscr{E}_i$ touches $\mathscr{E}$ at A and $\mathscr{E}'$ at A', and $\mathscr{E}_{i+1}$ touches $\mathscr{E}$ at B and $\mathscr{E}'$ at B'. Then tangent at T to $\mathscr{E}_i$, $\mathscr{E}_{i+1}$ passes through V, as do the lines AA', BB', and $(VT)^2 = (VA)(VA') = (VB)(VB')$.

1.14. Take point $\mathscr{C}_1$ as centre of inversion. There are two tangents from point $\mathscr{C}_2'$ to circle $\mathscr{C}_3'$, and these arise from the circles sought.

1.15. Take point of contact of $\mathscr{C}_2$, $\mathscr{C}_3$ as centre of inversion, then $\mathscr{C}_2'$, $\mathscr{C}_3'$ are parallel lines, and we seek circles through point $\mathscr{C}_1'$ which touch these lines. One such circle is the line through $\mathscr{C}_1'$ parallel to $\mathscr{C}_2'$, and this arises from solution touching $\mathscr{C}_2$, $\mathscr{C}_3$ at point of contact.

1.16. Take point of contact of $\mathscr{C}_1$ and $\mathscr{C}_3$ as centre of inversion. Then $\mathscr{C}_1'$ and $\mathscr{C}_3'$ are parallel lines, and $\mathscr{C}_2'$ is a circle touching $\mathscr{C}_1'$. One solution is the line parallel to $\mathscr{C}_1'$ which touches $\mathscr{C}_2'$, and there are three circles which touch $\mathscr{C}_1'$, $\mathscr{C}_2'$ and $\mathscr{C}_3'$.

1.17. Take point of contact of $\mathscr{C}_1$ and $\mathscr{C}_2$ as centre of inversion. $\mathscr{C}_1'$ and $\mathscr{C}_2'$ are parallel lines, and $\mathscr{C}_3'$ is a circle which touches neither. Two solutions are tangents to $\mathscr{C}_3'$ which are parallel to $\mathscr{C}_1'$.

1.18. Invert in $\mathscr{D}$. Then circle $\mathscr{C}$ becomes the line BC, and centre of $\mathscr{C}$ inverts into the point O' which is the inverse of the centre of $\mathscr{D}$, the point A, in the inverse of $\mathscr{C}$, which is the line BC. The inverse of O' in $\mathscr{D}$ is the centre O of the circle $\mathscr{C}$.

1.19. The circle LAM remains fixed as A moves. If AU is parallel to BC, meeting this circle again in U, show that U is fixed on this circle. It is the centre of the envelope of BC.

1.20. Reduce this case to the one treated above by drawing a parallel to AB to pass through the centre of $\mathscr{C}$, and a parallel to AC to pass through the centre of $\mathscr{D}$.

CHAPTER II

2.1. The circle $x^2 + y^2 - 2px - 2qy + r = 0$ is mapped on (p,q,r). If $r = 0$ the circle passes through $(0,0)$. The centres (p,q) of the circles lie on l, and the reflection of $(0,0)$ in the line l is the other common point of intersection. If l passes through $(0,0)$, the circles all touch at $(0,0)$.

2.2. The two circles are mapped onto the points $(p,q,-k^2)$, $(p',q',-k^2)$, and the line joining these points is parallel to the Oxy-plane and below Ω, so that if $k \neq 0$ it does not intersect Ω. Hence its polar line intersects Ω in two distinct points, and the coaxal system is of the intersecting type, with two distinct points of intersection.

2.3. If the parallel lines in E_3 are $(x - a)/l = (y - b)/m = (z - c)/n = t$, and $(x - a')/l = (y - b')/m = (z - c')/n = t'$, the circles are

$$x^2 + y^2 - 2(a + lt)x - 2(b + mt)y + c + nt = 0,$$
$$x^2 + y^2 - 2(a' + lt')x - 2(b' + mt')y + c' + nt' = 0,$$

for varying t and t', and each system has the radical axis:

$$2lx + 2my - n = 0.$$

2.4. Common chord of the two circles is:

$$(x - x')(a/c + 2x') + (y - y')(b/c + 2y') = 0.$$

2.5. Circles are orthogonal if their representative points in E_3 are conjugate. The point (p,q,r) is in the given plane, and therefore conjugate to the pole of the given plane.

2.6. Three points in E_3 which are not collinear define a plane with a unique pole with respect to Ω. The pole represents $\mathscr{C}_0$, the unique circle orthogonal to the three given circles. If the three circles are in a coaxal system of the intersecting type, the point-circles of intersection are the only contact circles. If non-intersecting, there are no contact circles.

2.7. Since $\mathscr{C}_i$ is orthogonal to $\mathscr{C}_0$, its inverse in $\mathscr{C}_0$ is itself. If we invert in $\mathscr{C}_0$ the three given circles invert into themselves, and contact circles invert into contact circles, so that the eight contact circles are paired.

Chapter III

3.1. The equation of the transformation which has (a,a'), (p,p), (q,q), and (z,w) as corresponding pairs of points is given by (2), p. 50, as:

$$(wa',pq) = (za,pq)$$

which leads immediately to the required form.

3.2. Since the cross-ratio of four points on a circle is real, we have $(za,bc) = t$, where a, b and c are fixed points on the circle, z a variable point, and t is real. From $(z - b)/(z - c) = t(a - b)/(a - c)$, we find z in the required form. If $AD - BC = 0$, we should find that $At + B = k(Ct + D)$, so that $z = k$, whereas z is a variable point on the circle.

3.3. Since the circular arc BC is orthogonal to ω the centre of the arc lies outside Ω, and therefore the arc BC is convex to the centre of ω.

3.4. The arcs AB and AC both pass through the inverse A' of A in ω. Take A' as centre of inversion, and for circle of inversion the circle Σ centre A' orthogonal to ω. Then ω inverts into itself, and since circles through A' and A invert into lines orthogonal to ω through the inverse of A in Σ, this inverse is the centre of ω.

3.5. If the h-line intersects ω in B and C, after an inversion which maps ω onto itself, B onto B' and C onto C', and A onto the centre O of ω, the inverse of the h-line BC is the h-line $B'C'$, and the Euclidean line through O and the centre Q of the circle defined by the arc $B'C'$ is perpendicular to the h-line $B'C'$, the lines joining the point O to B' and C' respectively touch the h-line $B'C'$, and OQ bisects the angle $B'OC'$.

3.6. If the given h-line is l, and the perpendicular h-lines are m and n, l and ω define a coaxal system of the intersecting type, and m and n are circles in the conjugate coaxal system, which is non-intersecting.

3.7. Two distinct h-lines with no intersection define a non-intersecting coaxal system, and have limiting points L and L' which lie on ω, since this is a member of the conjugate coaxal system. There is a unique circle through L and L' which is orthogonal to ω, and this is the unique h-line which is perpendicular to the given h-lines.